NF3, SF6 and the Apocalyptic Super Greenhouse Gases

R. Roy Blake

Introduction

As the wildfires that have plagued Canada, Europe and the western (and lately even the eastern) US, the increasing sea level threats to Pacific islands coastal areas, the shrinking polar and glacial ice, the increasing severity of hurricanes and floods, droughts and the unbroken continuation of "hottest months/years" all have demonstrated, climate change, global warming is an undeniable reality.

Also undeniable is that, to date, mankind's efforts to control and reverse climate change clearly have not been successful.

To date most of our efforts to alleviate climate change have been focused on carbon dioxide, (CO_2). With the exception of water vapor, CO_2 is the most benign of the greenhouse gases. In addition to energy production, CO_2 is emitted by animals including humans, volcanoes and other natural processes. CO_2 emissions also have the advantage in that they are naturally sequestered by the many CO_2 "sinks" of the natural world, including the ocean and all plant life.

Conversely, other greenhouse gases are many times more potent global warming agents than CO_2. These include methane, which is 28 - 90 times (depending on the time frame) more potent a greenhouse gas than CO_2, nitrous oxide, which is 300 times more potent.

Most menacing of all, are a group of greenhouse gases known informally as super-greenhouse gases.

One of those gases, nitrogen trifluoride (NF3), is estimated to be 17,200 times as potent a greenhouse gas than CO2. Another of the super greenhouse gases, Sulfur Hexafluoride (SF6) is over 20,000 times more potent a greenhouse gas than CO2.

Ironically, in addition to its use in the manufacture of computers and flat screen televisions, NF3 is used in the manufacture of solar panels. NF3, Nitrogen trifluoride, is a gas also used in the manufacture of thin film photovoltaic cells, microcircuits, and liquid crystal flat panel displays.

Worse, the use of NF3 is estimated to be increasing at rate of between 10 and 11% per year. At that rate, NF3 may overtake CO2 as the primary greenhouse gas by 2033. It does not take an expert to realize that if global warming increases by thousands, human survival, indeed the survival of all non-microscopic animal and plant life is extremely unlikely.

It may well be that much of the reason global warming appears to be accelerating at a pace much greater than predicted is already attributable, at least to some extent, to the super-greenhouse gases, in addition to increasing levels of methane and nitrous oxide. The greenhouse gas "mix" is most definitely changing. Nothing demonstrates just how fast the mixture of greenhouse gases is changing in favor of stronger gases than CO2, than the pre-COP29 statement of the "Climate Czar," that CO2 might only be responsible for slightly greater than 50% of total greenhouse gases. Just two years earlier the EPA claimed that CO2 was responsible for at least three quarters of greenhouse gases

As I get older I often lose track of where and when I first heard something. What I had heard is that a gas used in the production of flat screen TVs is a greenhouse gas 29,000 times more potent than carbon dioxide. While, through research, I was to find that while the 29,000 figure was an exaggeration, it was not a truly meaningful one.

Indeed, rather than focusing on the two numbers my first reaction was to question why a matter so important to the future survival of life on earth was not more well known. Followed by why isn't anybody doing anything about this, or is anyone doing anything about this?

NF3 is created by the reaction of hydrochloric acid (HCl), and ammonium (NH3). According to Professor F.J. Martin Torres at the University of Aberdeen in the UK, NF3 "…is toxic, odorless, colorless, and is an oxidizing gaseous at room temperature and atmospheric pressure…"

In a mixture of a break from the conventional narrative and understatement, in 2008 Science Daily called NF3 a "super" greenhouse gas and climate time bomb that could end up having "a greater impact on global warming than the world's largest coal-fired power plants."

Except for a protocol adding NF3 as a greenhouse gas that ought to be reported in an inventory, NF3 is not regulated by any environment treaty and can remain in the atmosphere for up to 550 years.

A study conducted at the University of California at San Diego revealed that atmospheric concentrations of NF3 were much higher than had been predicted at least in part due to underestimation of industrial losses of the gas.

The research was published in the October 31, 2008 edition of Geophysical Research Letters, a journal of the American Geophysical Union (AGU). The study revealed that instead of the estimated 2% of NF3 released to the atmosphere the actual figure was an alarming 16%.

Even worse, likely due to the increasing popularity of flat screen televisions, solar panels, etc., NF3 emissions are rapidly increasing. Emissions are estimated to have increased 40 fold between 1992 and 2007. One estimate has NF3 emissions increasing at the rate of 11% per year.

Add to that, in the words of Professor Martin-Torres of the University of Aberdeen: We have no reliable estimates about leakage (of NF3) during production, shipping and decommission. "This gas (NF3) is extremely volatile and **difficult to measure at low abundances**, so we cannot be sure that industry estimates or measurements are accurate."

Initially, NF3 was seen as an improvement on a greenhouse gas and ozone depleting gas called hexafluoromethane (C2F6) that was "controlled" by the Kyoto Protocol, the 1985 Vienna Convention for the Protection of the Ozone Layer and 1987 Montreal Protocol on Substances that Deplete the Ozone Layer. NF3 (and a myriad of other super greenhouse gases) was not addressed by the Kyoto Protocol or any other international environmental treaty.

As a result according to Professor Martin-Torres,

> "...NF3 became a strong competitor due to its lower costs and it initial absence from the Kyoto Protocol, NF3 'has a greater impact on Earth's climate per unit mass of emissions' than the C2F6 that it replaced..."

In addition to its lower cost and initial absence from the Kyoto Protocol Professor Martin-Torres cites additional NF3 advantages,

"NF3 offers a number of advantages of relative ease of use at ambient conditions that have made it very popular in industrial applications: it is non-corrosive, non-reactive and easy to manage chemical and reactivity hazards during storage, transportation, and normal operations."

"It has the ability to act as a stable fluorinating agent and has a wide application scope in high energy laser to dry etching in semiconductor production as a filling gas in lamps to prolong their durability and increase brightness, as well as a detergent gas in CVD apparatus. For all these reasons, nitrogen trifluoride is increasingly used in the electronics industry, primarily for the etching of microcircuits, and for manufacturing of liquid crystal flat panel displays and thin film PV (photovoltaic) cells."

Alarmingly, Professor Martin-Torres goes on to report that:

"The robust development of the consumer electronics market has produced a tremendous growth in the global nitrogen trifluoride market in the last years. In addition due to the rising demand for flat panel displays, LCD televisions and several other electronics products, the market for NF3 has exploded across the globe. In fact, NF3 production has consequently increased 15-

17% a year, from 1,000 tons…to more than 28,500 in 2019."

Asia is currently the largest regional NF3 market (responsible for 88% of NF3 emissions or 25,200 tons in 2019).." North America is responsible for 8.5% of NF3 emissions (2,400 tons). and Europe emitted 1,030 tons or 3.5% of the worldwide total.

The reason behind the industry's demand for NF3 is that manufacturers in the electronics industry all use a vacuum chamber to etch intricate circuitry and to deposit a thin layer of chemical vapor on the surface of a product. Some of the vapor inevitably builds up instead as a glassy nuisance on the interior of the chamber.

To tear apart that layer of crud and clean the vacuum chamber, manufacturers were using powerful fluorinated greenhouse gases. The usual choice, hexafluorethane, or C_2F_6, at first seems better than NF3. In terms of global warming it is 12,000 times more potent than CO_2, as opposed to the 17,200 times greater greenhouse gas potency of NF3.

However, approximately 60% of C_2F_6 ends up in the atmosphere. Initial estimates that proved far too optimistic were that, given optimal conditions, only about 2% of NF3 is released to the atmosphere. Of course, these estimates were proven to be wildly optimistic by subsequent studies.

When the semiconductor industry entered into a voluntary partnership with the EPA to reduce greenhouse gas emissions by 10% from 1995 levels between 1999 and 2010, NF3 became the preferred alternative.

Still more oddly, the world's largest producer of NF3 received a 2002 Climate Protection Award from the Environmental Protection Agency. In 2002, the EPA gave a Climate Protection Award to Pennsylvania based Air Products and Chemicals, Inc. **the largest NF3 producer.**

https://e360.yale.edu/features/the_greenhouse_gasthat_nobody_knew#:~:text=A%20practical%20alternative%20to%20NF3,on%20site%20using%20special%20equipment.

Considering the apocalyptic potentials of both NF3 and C2F6, a choice limited to just these two chemicals is depressing. Fortunately, there is at least one alternative to the use of either gas. According to Paul Stockman of Munich based Linde Gas, fluorine has zero global warming potential and no atmospheric lifetime.

Currently, the largest problem is that it is highly toxic and reactive. As it cannot be shipped in bottles as with NF3, fluorine must be generated on site using special equipment.

NF3 is not the only "super greenhouse gas" currently in use. For example, sulfur hexafluoride (SF6), is the only "super greenhouse gas" with a more detrimental environmental effect than NF3.

Sulphur Hexafluoride (SF6) is estimated to be 22,000-23,500 times greater a greenhouse gas than CO2. SF6 is used primarily in the production of transformers and magnesium.

It would be a mistake for us to view the threat posed by the "super greenhouse gases" as something that will only threaten us in the future. In fact, there is significant evidence that a combination of gases, including some of the "super greenhouse gases" are already worsening global warming.

Within the conclusion is a sample letter that readers who are as alarmed as myself about the super-greenhouse gases, and the lack of attention given to them are encouraged to use the letter or any portion to contact their lawmakers, media, environmental groups and other NGOs.

The book relies on statistics and other information culled from EPA, NOAA, NASA and other websites and scientific papers, as well as independent research, the most important of which was conducted at the University of California, Irvine and University of California at San Diego.

UPDATE

The original version of this book entitled "Galloping Towards Oblivion: *NF3, SF6 and the Apocalyptic Super Greenhouse Gases*, came out just prior to the 2024 COP29 UN Climate Conference in Baku Azerbaijan. It had been my hope that the book might assist me in making my case that perhaps the most intelligent strategy for combating global warming might not be to direct virtually all of our efforts addressing the most benign and naturally occurring of the greenhouse gases, while ignoring gases thousands of times worse.

Worse still, were our efforts to address the least harmful greenhouse gas while not addressing the "super greenhouse gases," actually worsening the problem, e.g., manufacturing solar panels employing the use of a greenhouse gas thousands of times greater than the CO2 the panels were attempting to address.

I did derive some hope (albeit incomplete) from the pre-COP29 statement from the Climate Czar office of the US State Department concerning "super pollutants."

In a paper called "COP29 Super Pollutants" the State Department (home of the "Climate Czar') released the shocking estimate that:

> "Super pollutants – non-CO2 greenhouse gases like methane, nitrous oxide and

hydrofluorocarbons (HFCs)---- are responsible for approximately half of near term global warming...."

I was at first taken aback by the Climate Czar's assessment that approximately half of current global warming was from non-CO_2 sources. That assessment was well at odds with the EPA's 2022 estimate that non-CO_2 gases contributed just 20-25% to total global warming.

Happily, the Climate Czar acknowledged that addressing greenhouse gases more powerful than CO_2 would have a more substantive effect addressing climate change than is currently the case. Unhappily, however, the Climate Czar's office only suggested action primarily addressing methane and nitrous oxide, gases it labelled "super pollutants."

While both methane and nitrous oxide are much more powerful greenhouse gases than CO_2 (methane is from 35 to 90 times more powerful and nitrous oxide is up to 300 times), they can hardly be considered "super" when compared to nitrogen trifluoride (NF_3 which is 17,200 times as powerful a greenhouse gas as CO_2)) or sulfur hexafluoride (SF_6), the only greenhouse gas more potent than NF_3, estimated to be over 20,000 times stronger than CO_2.

That State Department Climate Czar's assessment that non-CO2 greenhouse gases currently account for nearly 50% of total greenhouse gases directly and radically contradicts Environmental Protection Agency (EPA) assessments that CO2 alone is responsible for no less than 76% of total greenhouse gases.

The discrepancy points either to enormous uncertainty in greenhouse gas assessments or a rapidly changing situation that could easily get out of hand. There is very little margin for error here.

https://www.state.gov/cop29-releases-super-pollutants/

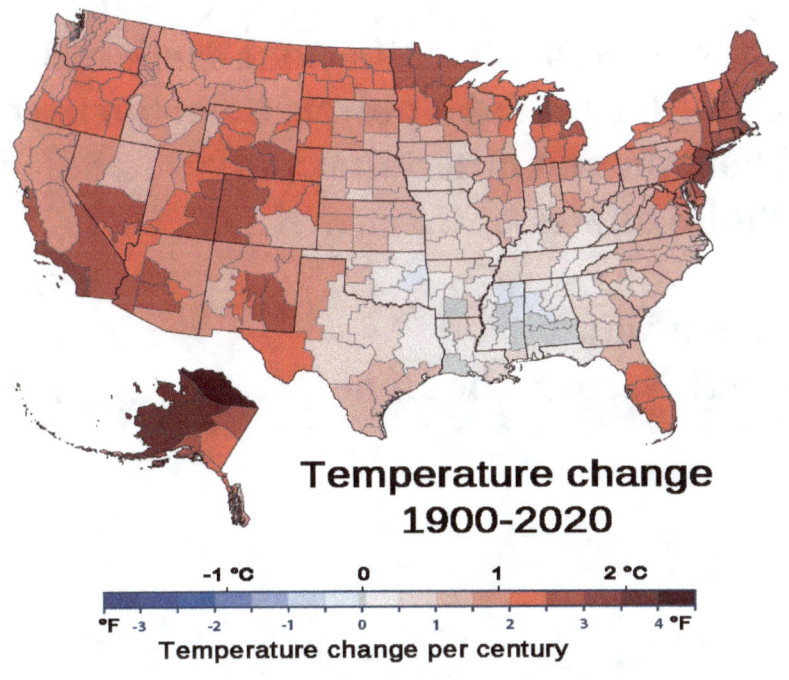

Temperature change 1900-2020

Table of Contents

Chapter One: Greenhouse Gases, "Regulated" and Unregulated………..19

Chapter Two: Is Global Warming Proceeding More Quickly Than Predicted?..37

Chapter Three: Questions Presented (the Big One)…………………...………40

Chapter Four: Geothermal, Helium 3, Landfill gas, Tidal power, Solar furnace, Thorium and other alternatives to Solar Photovoltaics………..……………………57

**Conclusion:
What Needs to be Done**…………..……...73

Appendices..................…................84

Appendix A

Percentages of volume of individual greenhouse gases to the overall greenhouse gas total volume per EPA 2022...88

Appendix B

Could Super Greenhouse Gases be Used to Terraform Mars?................................89

Appendix C

Increases in Global and East Asian Nitrogen Trifluoride (NF_3) Emissions Inferred from Atmospheric Observations.............................98

Appendix D

From the European Commission, Climate-friendly alternatives to HFCs…..…………….…..120

Appendix E

Total U.S. CO2 Greenhouse Gas Emissions in 2022…………...………..126

Appendix F

UC Irvine 10/29/2008 report on NF3………..……………….128

Appendix G

From Science Direct, Carbon Tetrafluoride, based on Chemical Engineering Journal, 2021…..………131

Appendix H

Email query to Sierra Club, EPA, NOAA, University of California, Irvine, climate scientists, University of San Diego, climate scientists and UN Environmental Programme, responses and follow-ups..............................151

Appendix I

Fighting global warming by GHG removal: Destroying CFCs and HCFCs in solar-wind power plant hybrids producing renewable energy with no-intermittency from International Journal of Greenhouse Gas Control, Volume 49....................................185

Appendix J

Best Practices to Reduce SF_6 Emissions..............…................193

Appendix K

**What is Nitrogen Trifluoride (NF3)
Professor F. J. Martin-Torres
Chaired Professor in Planetary Sciences
University of Aberdeen, Scotland
UK**………………..……………………198

Appendix L

COP29 Super Pollutants

(statement of the Climate Czar via the State Department to the COP29 UN Climate Conference 2024)…………..…206

Appendix M

The Sprint to Cut Climate-Super-Pollutants COP29 Summit on Methane and Non-CO2 GHGS…………..…………207

Appendix N

COP28: What Was Achieved and What Happens Next?………………………….21

Smoke from wildfires blocking sunlight from the San Francisco Bay Area on September 9, 2020

NASA photo

Chapter One:

Greenhouse Gases, "Regulated" and Unregulated

Currently there are six greenhouse gases that are currently regulated by the Kyoto Protocol.

Those are carbon dioxide, methane, nitrous oxide, hydrofluorocarbons, PFCs and sulfur hexafluoride.

In terms of the most current assessment of the relative contributions of greenhouse gases to the atmosphere the EPA in 2022 finds that carbon dioxide produces by far the most significant amount, 79.7% of the total.

According to the EPA, methane is responsible for 11.1%. nitrous oxide is responsible for 6.1% The combination of NF3, hydrofluorocarbons (HFCs), perfluorocarbons (PFCs) and sulfur hexafluoride (SF6) is 3.1%.

Carbon Dioxide

Carbon Dioxide (CO2) is primarily produced by the burning of fossil fuels, (coal, natural gas and oil), solid waste, other biological materials. Additionally, CO2 is produced as the result of some chemical reactions, such as from cement production. CO2 can remain in the atmosphere from 300 to 1000 years.

According to the 2022 EPA survey (albeit incomplete) of greenhouse gases, the annual total U.S. emissions of CO2 was 6,343 metric tons. The EPA estimates that the burning of fossil fuels in transportation was responsible for 35% of that total, the largest single proportion.

The second largest proportion (30%) was from the production of electricity. The EPA notes that of fossil fuels used in electrical production coal produced more CO2 than natural gas or oil.

The 2022 annual report from the National Oceanic and Atmospheric Administration (NOAA) stated that the global average atmospheric CO2 was 419.3 parts per million, which was a record high. The increase from 2022 to 2023 was 2.8 parts per million, the 12th year in a row in which the CO2 content in the atmosphere increased. The average annual atmospheric CO2 in 2023 was 421.08 parts per million.

https://www.climate.gov/news-features/understanding-climate/climate-change-atmospheric-carbon-dioxide

The EPA calls mitigations of greenhouse gases "sinks." In 2022 the most significant "sink" reported was "Land Use, Land Use Change and Forestry in the US" which resulted in a 13% offset of US produced greenhouse gases.

While some suspect that the 13% figure is too optimistic, even that figure is dwarfed by the at least 87% of greenhouse CO2 that was not offset.

Methane

Methane (CH4) is 28 - 90 times a more potent greenhouse gas over 100 years than CO2 and 84 times more potent over 20 years. Methane is also produced during the production and transportation of coal, natural gas and oil. Additional methane results from livestock production, other agricultural practices, land use, and the decay of organic waste, especially from landfills.

In 2022 the EPA estimated that methane was responsible for 16% of total greenhouse gases. NOAA lists it as one of the top five greenhouse gases.

In March of 2024 a non-binding "Global Methane Pledge" was issued at the UN Climate Change Conference in Glasgow at the COP26 conference by the US and the EU and 158 others.

In addition to roughly doubling the EPA's estimate of the contribution of methane to the greenhouse gas total, the pledge's webpage declares that:

https://www.globalmethanepledge.org/

> "Methane is a powerful but short-lived climate pollutant that accounts for a third (more than twice the EPA's 2022 estimate) of net warming since the Industrial Revolution."

While the pledge was signed by 111 nations including the United States and European Union, the total participants are only responsible for a minority (45%) of the planet's methane emissions. All of which makes the pledge's goal of reducing methane emissions by at least 30% below 2020 levels seem unlikely.

Nitrous Oxide

Nitrous oxide (N2O) molecules stay in the atmosphere an average of 121 years. It is 265 times a more potent greenhouse gas than carbon dioxide. N2O is produced by agricultural, industrial, land use practices, fossil fuel combustion, solid waste and wastewater treatment. Nitrous oxide has also been implicated with ozone depletion.

Unlike carbon dioxide, there are currently no practical methods for sequestering (removing) nitrous oxide from the atmosphere.

The greatest single source of nitrous oxide emissions comes from the use of nitrogen fertilizer. An alternative to the use of nitrogen fertilizers would be to plant crops that return nitrogen to the soil, such as legumes as a winter cover crop. Farmprogress.com points out, however, that the practice can deplete soil moisture.

Animal manure, compost, blood meal, grass clippings, wood ash, alfalfa meal, food waste such as eggshells and banana peels, yeast brew, drip irrigation and vermicompost also present alternatives to the use of nitrogen fertilizer.

The EPA's 2023 pie chart of greenhouse gases list nitrous oxide's addition to the total of greenhouse gases as 6.1, while at the same time listing CO2's contribution as almost 80%. Clearly, however, if the metric ton measurements of the two gases "carbon dioxide equivalents," the 6.1% figure cannot be correct.

If we take the actual metric ton measurements as correct, then the pie chart ought to have shown nitrous oxide emissions as roughly half those for CO2 or, over 28% instead of 6.1% and reduced CO2 from almost 80% to just 52%. Adjusted figures for methane and the "super greenhouse gases," might then reduce CO2's contribution to less than 50% of the total.

NOAA states that nitrous oxide emissions have grown 40% between 1980 and 2020, roughly echoing the estimate of a 1% annual increase in total greenhouse gas volume.

Nitrogen TriFluoride (NF3)

As noted earlier, NF3, nitrogen trifluoride, is 17,200 times more powerful a greenhouse gas than carbon dioxide (CO2). As also noted, NF3 can remain in the atmosphere for up to 550 years
Global NF3 emissions increased from 1.93 gigagrams per year in 2015 to 3.38 gigagrams per year in 2022, an average increase of approximately 10% - 11% per year.

The nations with the greatest use of NF3 are primarily in east Asia, although some is produced in the West.

The European Commission has issued an article entitled "Climate Friendly Alternatives to Fluorinated Greenhouse Gases" that speaks to NF3.

https://climate.ec.europa.eu/eu-action/fluorinated-greenhouse-gases/climate-friendly-alternatives-hfcs_en

Hydrofluorocarbons, perfluorocarbons, sulfur hexafluoride (SF6) and NF3 are lumped together in the EPA survey as being

> "synthetic, powerful greenhouse gases that are emitted from a variety of household, commercial and industrial applications...Fluorinated gases (especially hydrofluorocarbons) are sometimes used as substitutes for stratospheric ozone depleting substances."

As mentioned in the introduction Paul Stockman of Munich, Germany based Linde Gas, claims that there is at least one alternative to NF3. Fluorine, he states, has zero global warming effect and no lifetime in the atmosphere. However, it is highly toxic and reactive and must be generated onsite using special equipment.

Linde is promoting fluorine as an alternative to nitrogen trifluoride (NF3) after atmospheric scientists expressed concern that emissions of NF3 gas may contribute to global warming. Fluorine gas, which has zero global warming potential, can be substituted for NF3 in the production of electronic products including flat-panel displays, asserts Linde. The firm says customers such as Samsung and LG use Linde on-site fluorine generators. Air Products & Chemicals is the leading manufacturer of NF3.

While some NF3 manufacturers insist that there is no alternative to the use of NF3, the primary use of NF3 is to dissolve and remove silicon, silicon dioxide, etc. defects. While silicon, silicon dioxide, etc., resists many substances there are substances other than NF3 that will dissolve it. Examples include hydrofluoric acid (HF) and strong bases such as hot potassium hydroxide and hot sodium hydroxide solutions.

Sodium dioxide (Na02)has the doubly beneficial side effect of carbon capture. HF is not explicitly listed as a greenhouse gas but is considered to be a "contributor" to the greenhouse effect due to its potential for forming more complex fluorinated such as HFCs.

As noted earlier, apparently at least one substitute for NF3 and is currently being employed by Samsung and LG, i.e., the use of the non-greenhouse, though somewhat problematic, fluorine gas.

Hexafluoroethane (C2F6)

Hexafluoroethane (C2F6) has a global warming potential (GWP) of 9,200 meaning, of course, that it has 9,200 times the global warming potential as the same volume of carbon dioxide (CO2). Some estimates have, however, have been as high as 11,000. It can last an estimated 1000-10,000 years in the atmosphere.

Hexafluoroethane is emitted as a byproduct of industrial processes, including aluminum and rare earth's smelting, semiconductor manufacturing and flat panel display manufacturing.

Hexafluoroethane is also known as PFC-116, one of the most abundant perfluorocarbons (PFCs) in the atmosphere. In addition to tetrafluoromethane it is one of the most abundant PFCs in the atmosphere.

While PFCs are regulated under the Kyoto Protocol of the United Nations Framework Convention on Climate Change, few, if any concrete steps have actually been undertaken to control their atmospheric releases.

China is the primary global source of hexafluoroethane emissions. Between 2008 and 2019 China was responsible for nearly two thirds of global emissions.

The potency of PFCs is more difficult to assess due to what the EPA calls its positive radiative forcing effect which give PFCs a wide range of global warming potentials (GWPs) from the hundreds to the thousands.

Hexafluoroethane is one of the two most abundant perfluorocarbons, (PFCs) in the atmosphere.

According to the European Union there are alternatives (although there is no "one size fits all" alternatives to hexafluoroethane and the other HFC gases.

https://climate.ec.europa.eu/eu-action/fluorinated-greenhouse-gases/climate-friendly-alternatives-hfcs_en

> **"To avoid the use and emissions of hydrofluorocarbons (HFCs), a variety of climate-friendly, energy-efficient, safe and proven alternatives are available**.
>
> Due to different thermodynamic and safety properties of the alternatives, there is no 'one size fits all' solution. The suitability of a certain alternative must be considered separately for each category of product and equipment; and in some cases also take into account the geographical location in which the product and equipment are used.
>
> The climate impact of a substance is commonly expressed as the global warming potential (GWP). The lower the GWP, the more climate-friendly the substance.
>
> HFCs have a very high GWP and are hence potent greenhouse gases. Most of the HFCs are used as refrigerants in refrigeration and air

conditioning (RAC) equipment, but also as blowing agents, aerosol propellants and solvents.

To mitigate emissions of substances with a high GWP and comply with the F Gas registration, each sector needs to find solutions to quickly switch to low GWP refrigerants.

Most of the HFCs are used as refrigerants in refrigeration and air conditioning (RAC) equipment, but also as blowing agents, aerosol propellants and solvents.

In the following, alternatives to commonly used HFCs are listed for different sectors.

The alternatives include:

    Natural refrigerants

    HFCs with lower GWP, such as R32

    Hydrofluoroolefins (HFOs)

    HFC-HFO blends."

Other alternatives include propane and isobutene, ammonia (NH3), carbon dioxide (CO2) and certain hydrofluoroolefins (HFOs).

Carbontetrafluoride (CF4)

Carbontetrafluoride (CF4), sometimes written as carbon tetrafluoride, has a greenhouse warming potential estimated to be 5,700 to 6,500 times that of CO2. Due to its stability, it may remain the atmosphere for 10,000 to 50,000 years.

It is useful as a low temperature refrigerant, in electronics microfabrication alone or combined with oxygen as a plasma etchant for silicon, silicon dioxide, silicon nitride, chlorine production and neutron detectors. Most estimates attribute CF4 contribution to the total of all greenhouse gases to less than 1%.

Alternatives to CF4) include nitrogen gas (N2), oxygen gas (O2), and carbon dioxide (CO2), when used in specific applications with carbon capture, as well as HFCs with much lower GWPs.

Chlorotrifluoromethane (CF3Cl)

Chlorotrifluoromethane (CF3Cl), is between 10,800 and 14,000 a more powerful greenhouse gas than CO2. It is an existential threat to the ozone layer, and can remain in the atmosphere for 640 years.

Also known as CFC-13, R-13 or Freon 13, Chlorotrifluoromethane is used as a low temperature refrigerant, a coolant, for etching glass, metal hardening, for degreasing printed circuit boards, for dry cleaning textile, as a bulking agent for polymer foams and as an aerosol propellant.

More environmentally friendly alternatives, which is a potent ozone-depleting substance, while including super greenhouse gases such as HFCs and HCFCs, newer options include hydrofluoroolefins (HFOs) which have significantly lower GWPs. In some cases water-based cleaning solutions or other alternatives may suffice depending on the intended use.

Perflubutane (C_4F_{10})

Perflubutane (C_4F_{10}) which is used in fire extinguishers, ultrasound contrast and nuclear medicine is estimated to be 8600 more powerful a greenhouse gas than CO_2. Perflubutane can remain in the atmosphere for an estimated 2600 years.

C_4F_{10}'s newly found uses as an ultrasound contrast agent and in nuclear medicine currently complicate efforts to find alternatives.

Octafluropropane (C_3F_8)

Octafluropropane (C_3F_8), is estimated to be 8000 to 24,000 times more problematic a greenhouse gas than CO_2. Octafluropropane (C_3F_8), is used in semiconductor manufacturing, and remains in the atmosphere for 2600 years.

More environmentally friendly alternatives to C3F8 include propane, ammonia, carbon dioxide and some hydrofluoroolefins (HFOs)

Sulfur Hexafluoride (SF6)

Sulfur Hexafluoride (SF6) is the only "super greenhouse gas" with a more detrimental environmental effect than NF3. SF6 is estimated to be 22,000-23,500 times greater a greenhouse gas than CO2.

(SF6) is the only one of the "super" greenhouse gases regulated by the Kyoto protocol, and will be present in the atmosphere for an estimated 3200 years. It is used primarily in the production of transformers and magnesium. The EPA has arguably labelled **sulfur hexafluoride, (SF6)** as the most potent greenhouse gas known.

There are several alternatives to the straight use of SF6 gas including using purified air as an insulating gas. The biggest drawback is that clean air systems take up more space than SF6 systems.
Another technique for lessening the negative climatic impact of SF6 involves using vacuum interrupters, which are commercially available in medium voltage breakers, for switching functions.

A mixture of C4 gas reduces the global warming potential of SF6 by 99%. C4 has the same electrical ratings as SF6 and can operate in the same temperature range.

Another alternative is a mixture of 20% oxygen and 80% nitrogen synthetic air, mainly used in vacuum switching technology.

Other alternatives include stunt vacuum interruption which combines vacuum interruption with a disconnector in air and MV switchgear solutions, that use air and solid insulated designs.

Fluoroform (CHF3)

Fluoroform (CHF3), is estimated to be 12,000 times more powerful a greenhouse gas than CO2. (CHF3), is used as a refrigerant and in the semiconductor industry, can theoretically remain in the atmosphere for 260 years. CHF3 is generated as one of the main waste products in the preparation of Teflon. Potential new medical uses complicate efforts to seek alternatives.

Hydrofluorocarbons (HFCs)

Hydrofluorocarbons (HFCs), any of several organic compounds composed of hydrogen, fluorine and carbon in use primarily as refrigerants, are 14,800 times as potent a greenhouse gas as CO2 (also designated as its

Global Warming Potential or GWP) more potent a greenhouse gas than carbon dioxide over a 100-year period.

The National Oceanic and Atmospheric Administration's (NOAA) Global Monitoring Laboratory states that global concentrations of HFCs have increased "rapidly over time." NOAA attributes this to the "enhanced use of HFCs as substitutes for CFCs, HCFCs and other chemicals as solvent/cleaning agents, refrigerants, foam-blowing agents, air conditioning fluids, etc., beginning in the late 1980s and early 1990s.

Hydrofluorocarbons, perfluorocarbons, sulfur hexafluoride (SF6) and NF3 are lumped together in the EPA survey as being "synthetic, powerful greenhouse gases that are emitted from a variety of household, commercial and industrial applications…Fluorinated gases (especially hydrofluorocarbons) are sometimes used as substitutes for stratospheric ozone depleting substances. NOAA lists HFCs as one of the top 5 greenhouse gases.

Alternatives to HFCs include natural refrigerants, Hydrofluoroolefins (HFOs), HFC-HFO blends, propane and isobutene, ammonia (NH3), carbon dioxide (CO2) and certain hydrofluoroolefins (HFOs).

Perfluorocarbons (PFCs)

The potency of perfluorocarbons (PFCs) as greenhouse gases are more difficult to assess due to what the EPA calls its positive radiative forcing effect which give PFCs a wide range of GWPs from the hundreds to the thousands.

PFCs are used in a number of household and commercial products as waterproofing agents, lubricants, sealants and leather conditioners.

Hydrochlorofluorocarbons (HCFCs)

HCFCs are estimated to be 2000 times more potent a greenhouse gas than carbon dioxide. Recent studies of atmospheric greenhouse gases have found that the presence of HCFCs have fallen by a modest three-quarters of one percent, as they are being phased out by international treaties per the Montreal Protocol which are aimed at preserving the ozone layer. HCFCs have one of the shortest survival rates in the atmosphere, estimated to be as little as 20 years. Still, NOAA lists HCFCs as one of the top five greenhouse gases.

Uses include as insulation materials and cooling agents. HCFCs were used as substitutes for CFC in insulation materials from 1991 through 1996, however, this use was prohibited in 1997.

Water Vapor, Ozone and others

Additional compounds that are considered greenhouse gases include water vapor, ground level ozone and other solid and liquid aerosols. In the initial stages of global warming, the atmosphere will retain more water vapor than had previously been the case.

According to the US Energy Information Agency

> "Ozone is technically a greenhouse gas, but is helpful or harmful depending on where it is found in the earth's atmosphere. Ozone occurs naturally at higher elevations in the atmosphere (the stratosphere where it forms a layer that blocks ultraviolet (UV) light which is harmful to plant and animal life, from reaching the earth's surface. The protective benefit of stratospheric ozone outweighs its contribution to the greenhouse effect and to global warming. However at lower elevations of the atmosphere (the troposphere), ozone is harmful to human health.
>
> There are human-made industrial chemicals that break down ozone in the stratosphere and create holes in the ozone layer. The United States and countries around the world ban and control and use of several of these industrial gases under the Montreal Protocol...."

Residents of low-lying Pacific Islanders watch as the high tides flood their homes and infrastructure

Chapter Two: Is Global Warming Proceeding More Quickly Than Predicted?

The simple answer is yes. Global warming is far outpacing early projections, although predictions have become increasingly dire over time and as one "hottest month on record" follows another.

July 2024 was the 14th consecutive "hottest month on record," in most of the world. The National Oceanic and Atmospheric Administration (NOAA) reported that

January through July 2024 temperatures were 2.3 degrees Fahrenheit (1.28 degrees Celsius) warmer than the 20th Century average.

As a result, projections of future global warming are being revised to better reflect the rapidly worsening situation. Climate scientist James Hansen now predicts that global temperatures will rise 2 degrees Celsius (3.6 degrees Fahrenheit) over pre-industrial levels by 2050.

That estimate is considerably more dire than estimates put forward by the Intergovernmental Panel on Climate Change (IPCC). Unfortunately, it would not be surprising if Dr. Hansen's predictions turn out to be too conservative, possibly far too conservative.

NASA reports that "…the summer of 2023 was Earth's hottest summer on record, 0.41 degrees Fahrenheit (F) (0.23 degrees Celsius (C) warmer than any other summer in NASA's record and 2.1 degrees F (1.2 C) warmer than the average summer between 1951 and 1980.

In his Ted Talk Environmental Science Professor Johan Rockstrom and Director of the Potsdam Institute for Climate Impact Research, readily concedes that climate change is clearly proceeding at a pace far in access of previous predictions. He even lists a number of "tipping points" that may already have been passed. One of the most alarming of these is that rather than being the greatest "carbon sink" on earth, the Amazon rainforest may already be or is on the verge of becoming a carbon dioxide source.

https://www.ted.com/talks/johan_rockstrom_the_tipping_points_of_climate_change_and_where_we_stand?subtitle=en

What is not as simple is assigning a cause or causes to the rapidly worsening situation.

Chapter Three: Questions Presented (the Big One)

NF3, is a chemical used in high tech "chamber cleaning," including in the production of solar panels, thin film solar photovoltaic cells, computers and, of course, flat screen televisions. NF3 also appears to be an incredibly powerful greenhouse gas, 17,200 times as potent as CO2. Further, while CO2 emissions appear to be roughly stabilizing, NF3 emissions are increasingly exponentially.

Given these facts, the question that then seems to most immediately (and importantly) to mind, is:

Would it be cheaper and/or less harmful to the environment to outlaw NF3 and simply adopt a more laissez-faire attitude towards more conventional greenhouse gases, or at least address the most dangerous greenhouse gases as (or more) aggressively than we are attempting to address the least dangerous greenhouse gas, carbon dioxide?

If for no other reason than the greenhouse gases already in the atmosphere will continue heating the planet at the current rate at least, the dismal truth appears to be that no matter which road we take, climate change seems inevitable for perhaps hundreds of years. That is certainly the case if we limit ourselves to addressing only certain greenhouse gases.

One of the difficulties in answering the question of the most intelligent course of action is the difficulty in quantifying it. Among the possible known quantities necessary for such a calculation are the total global CO_2 and NF_3 emissions, and the rate at which they are growing or stabilizing.

In 2008 global CO_2 emissions were 31.6 billion metric tons. In 2023 worldwide CO_2 emissions increased by 1.1%, or 410 million metric tons to 37.4 billion metric tons.

NF_3 appears to be increasing by at least 10% annually. In 2008 67 metric tons of NF_3 were released into the atmosphere, which given the 17,200 figure meant that the NF_3 emissions were the planet warming equivalent of 123,600 million metric tons of CO_2.

As the manufacture of solar panels is a primary source of NF_3 emissions, an additional variable might be an estimate of the average NF_3 released per solar panel. Unfortunately, currently that number can only be guessed at.

Following that an additional variable would have to be the efficiency of solar panels in replacing energy currently produced by CO_2 emitting energy production.

In an article by Aris Vourvoulias entitled "How Efficient are Solar Panels in the UK?" on Greenmatch, "typical average" residential systems in the UK are estimated to be only approximately 20% efficient, while more expensive systems "can be as much as 40% to 50% efficient.

Vourvoulias goes on to state that "Monocrustalline solar panels (15 – 22% efficient) are considered the best type, followed by polycrystalline (15 – 20% efficient) and thin-film solar panels (10 – 20% efficient), being as well the most common choice. Extremely hot climates can also reduce efficiency with the ideal panel temperature being around 25 degrees Celsius (77 degrees Fahrenheit)."

The most frequently cited solar panel efficiency, the rate at which sunlight is converted to electricity, varies between 15% and 22% for typical home systems…"

https://www.greenmatch.co.uk/blog/2014/11/how-efficient-are-solar-panels

While the less than 25% average solar panel efficiency seems unattractive given the great efforts and hopes tied to them, Vourvoulias's figures may, in fact, be too optimistic. During the early employment of solar farms by the Tennessee Valley Authority (TVA), the general efficiency was usually considerably less than 10%. In 2023 wind and solar combined were only responsible for 4% of TVA's power production.

Given that the production of solar panels involves the use of NF3, the suspicion that their production impacts the efforts to reduce global warming as, or perhaps even more negatively than positively, seems increasingly likely.

Currently, it takes 2000 500-watt solar panels to produce one megawatt of electricity. The typical coal fired plant generates roughly 1,000 megawatts. Gonghe Talatan Solar Park in China has a capacity of 15,600 megawatts. It is important to remember that the capacity of a solar (or wind) power plant is always its maximum capacity, an efficiency rarely, if ever, achieved. 1.8 billion solar panels were produced worldwide in 2022.

Of course, the question of importance to the survival of mankind is whether the production of solar panels (assuming the continued use of NF_3 at current and/or projected levels in their production) is eventually more contributory to climate change than the fossil fuel facilities that they replace. Unfortunately, an equation that would directly address that situation is elusive, as there is no agreed upon number for average NF_3 released by each solar panel. Still, much can be inferred.

The simplest equation would be to take the percentage of annual increase or decrease of both CO_2 and NF_3 and calculate the probable future, i.e., at what point, if any, that NF_3 will overtake CO_2 as a percentage of total greenhouse gases. In 2023 NOAA reported that the total presence of CO_2 in the atmosphere was 419.3

parts per million (ppm), an increase of 2.8 ppm over the previous year or an increase of 0.00067%.

At the same time NF3 emissions have been increasing at the rate of approximately 10% per year as per below link.

https://pubs.acs.org/doi/10.1021/acs.est.4c04507

Generously, fudging the percent of increased CO2 emissions to 1% from 0.00067% (in fact due to the substitution of natural gas generation for coal generation, CO2 emissions have actually gone down in the US and many European nations) using the much more accurate 10% per year to determine when NF3 emissions would constitute a greater percentage of greenhouse gases than CO2, projects that NF3 may well overtake CO2 possibly as early as 2033.

It doesn't take any sort of expert to realize that were a super-greenhouse gas or gases to become the primary component of global warming human survival seems very unlikely.

The unfortunate truth is that, short of miraculous intercession or the solution to the problem of cleansing our enormous atmosphere of greenhouse gases, global warming seems to be built into the system for at least the next 300 to 1000 years no matter what we do.

The question then becomes what is the most intelligent course of action we can take to minimize the all but inevitable abyss that we now face.

The way that we have apparently been answering that conundrum is to put almost all of our energies towards eliminating fossil fuel usage and thereby carbon dioxide. This is even when some of those efforts come with unforeseen consequences that complicate and potentially produce even worse outcomes.

In addition to being one of the weakest of greenhouse gases, of all the greenhouse gases (with the exception of water vapor), carbon dioxide is absolutely vital for life on earth. Of all the greenhouse gases, carbon dioxide has the most natural sinks (in the form of all plant life).

All of which begs the question, is carbon dioxide being unfairly or at least unwisely made the primary focus of efforts to control or at least alleviate global warming? Is there a scientific or at least mathematical effort to test that thesis? Can we create an equation that best models the situation?

A major difficulty in creating an equation that accurately models the situation is that often there are no generally agreed upon statistics in many cases. For example, while one government website predicts that NF_3 emissions will increase by 10%, another predicts that the increase will be 11%. Worse, still there is no general agreement on CO_2 annual growth, for which the estimates range from a net loss to a growth rate of 1%.

Another major difficulty is that there is a confusing mixing of units including metric tons, global warming potential (GWP), units of installed capacity, gigawatts, and CO2 equivalents, etc.

One of the chief variables we would need to attempt to solve any equation that would model our climate conundrum, is to determine the relative percentage of individual greenhouse gases historically and how they might be expected to increase or decrease in the future.

That might, therefore, reveal the most intelligent course of action, starting with the most recent data, working first backwards and then presuming future results, noting any additional variables that might impact future results.

Starting with the most recent data (which is so new that it is not official and can only be approximated), on Friday September 13, 2024 the Google AI Overview reported that:

> "According to the most current data, carbon dioxide (CO2) is the primary contributor to global warming, accounting for roughly 76% of total greenhouse gas emissions, followed by methane (CH4) at around 16% and nitrous oxide at 6%."

According to the EPA in 2015 CO2's contribution to the total was relatively unchanged at 76%, methane was also unchanged at 16% and nitrous oxide was also unchanged at 6%.

In the current preliminary report from AI, there was nothing at all about NF3, etc., suggesting that even AI is asleep at the wheel but we can still make some rough calculations by contrasting the most recent official data from 2022.

In 2022 the EPA reported finds that carbon dioxide produces by far the most significant amount, 79.7% of the total. According to the EPA, methane is responsible for 11.1%. nitrous oxide is responsible for 6.1% The combination of NF3, hydrofluorocarbons (HFCs), perfluorocarbons (PFCs) and sulfur hexafluoride (SF6) is 3.1%.

While the figures from 2015 through the preliminary figures for 2024 were remarkably similar, the changes from both in 2022, (as well as the wide range in the estimates of the effects of both methane and nitrous oxide) might lead a skeptic to wonder how well any of the figures can be trusted, not to mention the minimum 25% discrepancy between EPA and Climate Czar figures.

Be that as it may in less than two years the data then suggests that the overall contribution of carbon dioxide has decreased by more than 3%, while methane has increased almost by nearly 5%. Nitrous oxide has remained steady and, quite surprisingly NF3 and other sources appears to have dropped by about one percent.

If those figures are correct and do indeed portray an 8% shift over a two-year period, methane can be expected to overtake CO2 as a percentage of greenhouse gases by 2040. As methane is 28 - 90 times a potent greenhouse gas as CO2 that hardly seems like good news.

That decline might be best explained by the fact that the "most recent results" are all currently approximations and the almost shocking increase in methane emissions resulting in an increasing total of greenhouse gas emissions that might well have masked the fact that NF3, etc. emissions have either remained steady or, (more likely) have increased.

Further, in July 15, 2024 Environmental Science and Technology reported that NF3 emissions inferred from atmospheric observations have increased alarmingly from the 1970s to the present.

In 1975 the inferred presence of NF3 was nearly non-existent, a mere 0.008 parts per trillion (ppt). By 2011, however, the inferred NF3 had jumped to 0.86 parts per trillion (give or take 0.04 ppt). In 2020 NF3 was inferred at 2.3 ppt, jumping to 2.59 ppt (plus or minus 0.11 ppt) the following year.

While the relative percentage of greenhouse gases attributable to individual gases may decline while at the same time it is increasing due to the fact that overall amounts of all gases have increased.

In fact, the 2022 annual report from the National Oceanic and Atmospheric Administration (NOAA) the global average atmospheric CO_2 was 419.3 parts per million, which was a record high. The increase from 2022 to 2023 was 2.8 parts per million, a less than one percent increase. However, it was also the 12th year in a row in which the CO_2 content in the atmosphere increased.

The average annual atmospheric CO_2 in 2023 was 421.08 parts per million. While CO_2 emissions increased slightly, the percentage of their total contribution to all greenhouse gases apparently decreased if NOAA's preliminary 2024 findings are to be believed.

I then asked, Google's AI Overview if NF3 use has been increasing:

> **"Yes, the use of NF3 is significantly increasing**
>
> primarily due to its widespread application in the electronics industry for manufacturing processes like etching microcircuits in the production of flat panel displays, semiconductors and solar cells. This rising demand is causing a concerning

increase in NF3 emissions as it is a potent greenhouse gas."

According to "Increase in Global and East Asian Nitrogen trifluoride (NF3) emissions inferred from Atmospheric Observations" on MIT's Joint Program on the Science and Policy of Global Change:

"Global NF3 emissions have grown from 1.73 0.13 Gg yr in 2014 to 2.91 0.23 Gg yr in 2020."

That is a 68% increase over six years or more than 10% per year. While in 2015 the EPA reported that the total of all greenhouse gases increased 44% from 1990, a 25 year period. While hardly reassuring, that increase is less than one percent a year, a figure dwarfed by the more than 10% annual NF3 increases from 2014 through 2020.

However, another EPA report from 2020 stated that largely due to the shift to natural gas power generation from coal, total net emissions of all greenhouse gases had actually declined 21% from 2005. Although as the reason for the decline is almost due to a CO2 decline, the much more potent greenhouse gases have not experienced similar declines.

https://www.epa.gov/system/files/documents/2022-04/us-ghg-inventory-1990-2020-data-highlights.pdf

Other variables to possibly consider might be future developments including future sequestration of greenhouse gases, both in the atmosphere and otherwise, substitutions for individual greenhouse gases and (hopefully) other future efficiencies.

In terms of the sequestration of greenhouse gases, most efforts have been aimed at, the least dangerous of the greenhouse gases, CO2. Interestingly, Exxon Mobil claims to be the leader in the field of Carbon Capture and Storage, claiming to have cumulatively captured and sequestered 120 million metric tons, or 40% of all the carbon dioxide ever sequestered.

https://corporate.exxonmobil.com/what-we-do/delivering-industrial-solutions/carbon-capture-and-storage?utm_source=google&utm_medium=cpc&utm_campaign=1ECX_GAD_TRAF_OT_Non-Brand_Carbon+Capture&utm_content=OT_Non-Brand_Car

While 120 million metric tons may seem significant, on March 18, 2024 NASA reported that 1079 metric tons of CO2 are being released into the atmosphere. There are 31,536,000 seconds in a year and, if the NASA figures are reliable, 3.4 billion metric tons of carbon dioxide are released annually, dwarfing Exxon Mobil's (and the world's) heretofore success in CO2 sequestration. As noted earlier, fortunately, there are a lot of natural CO2 sinks, the largest of which are, of course, the oceans all plant life.

https://science.nasa.gov/science-research/earth-science/climate-science/how-much-carbon-dioxide-are-we-emitting/

Ironically, other than the exploitation of methane in landfills and other methane rich environments, methane sequestration involves turning it into CO2. Typically, methane sequestration is described as the use of oxidation by hydroxyl radicals through which methane molecules are broken down and converted into carbon dioxide.

In spite of the EPA and NOAA estimates that the contribution of methane towards the greenhouse gas total has either increased by 5% or remained steady, the American Petroleum Institute (API) contends that:

> "average methane emissions intensity declined by nearly 66 percent across all seven major onshore producing regions from 2011 to 2021."

In fact, due to the very specific terms used by the API, both reports may still be correct and not contradict one another.

https://www.api.org/news-policy-and-issues/methane?utm_source=AD&utm_medium=SEARCH&utm_campaign=AMPET-0044&utm_content=METHANE_CO&gad_source=1&

gclid=EAIaIQobChMI2uf45ofIiAMVp9XCBB2HMyR
bEAAYASAAEgIHmPD_BwE

Unfortunately, complicating any attempt to create an equation that would reveal the most intelligent strategy for addressing global warming is that the statistics have not, in many cases been consistent or in overall agreement with other statistics.

At this point, however, it should be clear that the strategy of focusing, almost exclusively on combatting carbon dioxide emissions, addresses the least existential threat to human survival of the greenhouse emissions.

While it is possible to conceive that the human race could, with difficulty, survive the rising CO_2 levels, that, unfortunately, are already inevitable, human survival of rising and extensive emissions ranging from 28 to 17,200 times as potent as CO_2 seems increasingly unlikely.

An equation that modeled the problem might be one that calculates the priority of addressing each greenhouse gas. The easiest way that might be done is to assign a numerical value to each factor resulting in a solution which can accurately portray the priority each gas requires with the largest number reflecting the highest priority.

One possible equation that might be helpful in modeling the problem could begin by assigning each greenhouse gas a value based on its current percentage contribution to the total of greenhouse gases.

Thus, CO_2 starts out with a big lead of 76 reflecting the value of its most recently measured contribution to the total of greenhouse gas and their effects of 76%. Methane comes in second at 16, nitrous oxide is 6.1 and NFE etc., is either 2 or 3 or somewhere in between. So that the we may proceed we will give it 2.5.

Percentage change over a given period could be added, subtracted or multiplied by the number of of the change in percentage over a given period of time.

The interim figures for 2024 compared to the official numbers for 2022 reveal a 1.5% per year decline for CO_2, and a 2.5% increase for methane. For this purpose, a less than 1% a year increase for nitrous oxide as well as for NF_3 despite the fact that releases of both have continued to increase.

CO_2 falls to 73, while methane rises to 21, nitrous oxide and NF_3 remain unchanged at 6.1 and 2.5 respectively.

Most significantly that number might then be multiplied by the potency of each greenhouse gas, i.e., one for CO_2, 38 for methane, nitrous oxide is 300 and NF_3 is 17,200.

CO2 remains 73, methane rises to 588, nitrous oxide is somewhere between 1830 and 1616.5 and NF3 rockets to 43,000.

Thus, according to this equation NF3 ought to be 589 times a greater priority than CO2. Nitrous oxide is 22 times more important and methane is eight times as important.

As stated earlier methane is projected to pass CO2 as the predominant greenhouse gas by 2040 even while total greenhouse gases continue to rise. Given the projected increases in NF3 releases of nearly 10% annually, a point in history in which NF3 becomes the primary greenhouse gas is almost too terrible to contemplate.

Finally, the percentages of each greenhouse gas by volume, does not tell the whole story, the most important story. The effects of NF3 and the other super greenhouse gases are not limited to the future. They are, in fact, already with us today and increasing with each passing year.

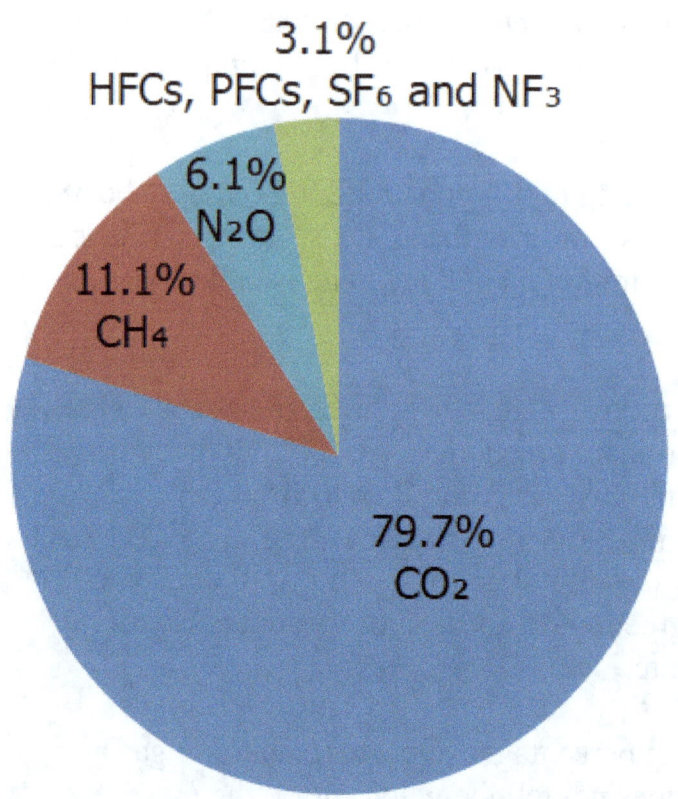

The above image is taken from the EPA report on greenhouse gases from 2022.

Ultimately, however, such an equation, especially a generally agreed upon equation does not seem possible at the present time. Certainly, however, the question of whether it is more environmentally friendly to support natural gas (which has replaced coal in much of the West) instead of solar panels produced using NF3 does seem to lean towards the older technology, if the choice is reduced to those two.

Chapter Four:

Geothermal, Helium 3, Landfill gas, Tidal power, Solar furnace, Thorium, Fusion, Cogeneration and other alternatives to Photovoltaics and Wind Energy

In the event that no alternative to the use of NF3 is practical in the production of solar photovoltaics, are there other alternative energies that can be employed that do not produce greenhouse gases?

Virtually the entire effort to produce electricity without attendant CO2 emissions, have been focused on wind and/or solar energy. It hardly bears repeating those alternatives are limited to the whims of nature, i.e., when the sun shines and/or the wind is blowing.

Geothermal

Yet there is a no shortage of other alternatives, some of which produce power constantly, the most well known of which is geothermal.

So, what is geothermal's environmental impact and how does it compare to that of either NF3 or CO2?

If we start by comparing the use of the rare earth element, for which there is an undeniable cost both in terms of their extraction and their processing of wind energy versus geothermal we are in for a pleasant surprise.

In a January 6, 2024 article at Science Direct, entitled "Rare Earth Element Extraction from Geothermal Brine Using Hybrid Capacitive Deionisation Process," we discover that unlike wind energy, which is heavily dependent on rare earth mining and processing, geothermal, in fact, has the potential to become a source of rare earth elements production. In the words of the article:

> "Geothermal fluids are potentially significant sources of valuable minerals and metals in the form of rare earth elements." The key word there is "potentially," which, of course, means no one is doing it.

https://www.sciencedirect.com/science/article/abs/pii/S221478532305335X#:~:text=Along%20with%20the%20heat%20energy,%2C%20sustainable%20system%20%5B2%5D.

On the other hand, it is of no shock that the environmental record involved with rare earth's extraction and processing is hardly comforting.

In 2020, REE mining emitted 152.97 kilotons of carbon dioxide, a 94% increase from 2010.

Hydrochloric acid is the biggest contributor to the total greenhouse gas footprint of REE processing accounting for approximately 38%.

Steam use accounts for 32%, electrical usage accounts for 12% of REE related greenhouse gases.

Worse, however, the extensive volume and radioactivity of REE mining tailings can contaminate food sources and cause respiratory issues. Villages near the Bayan Obo mine in China has experienced higher rates of cancer and respiratory illness.

What is the potential of geothermal and what are the pros and cons of its further development?

Where a substantial geothermal potential exists, often indicated by the presence of natural geothermal emissions, such as hot springs and volcanic activity, geothermal has a proven history of safe use.

Perhaps the most successful employment of geothermal power has been in Iceland. Located on the primary planetary engine of seafloor spreading and thus, plate tectonics, the mid-Atlantic ridge, geothermal accounts for a full two thirds of Iceland's energy needs.

According to Climate Care.com, the ten largest geothermal power plants in the world are all, with the exception of the Larderello complex in Italy, situated either along the mid-Atlantic ridge or the Pacific ocean "Ring of Fire:"

1. The Geysers Geothermal Complex, in California USA produces 900 megawatts of electrical energy. Historically the Geysers has provided sufficient electrical generation to power roughly 1,000,000 homes.

2. Larderello Geothermal Power Complex in Italy which produces 769 megawatts. Although Italy is not located on either the mid-Atlantic ridge or in the ring of fire, it has significant geothermal resources as evidenced by the still active and dangerous volcanoes of Mounts Etna, Stromboli and Vesuvius.

3. Cerro Prieto Geothermal Power Station in Mexico which produces 720 megawatts.

4. Makban Geothermal Complex in the Philippines which produces 458 megawatts

5. Calenergy Generations Salton Sea Geothermal Power Plants in California, USA 340 megawatts

6. Hellisheidi Geothermal Power Plant in Iceland which produces 303 megawatts.

7. Tiwi Geothermal Complex in the Philippines, which produces 289 megawatts.

8. Malitbog Geothermal Power Station in the Phillipines, which produces 232.5 megawatts.

9. Wayang Windu Geothermal in Indonesia which produces 227 megawatts.

10. Darajat Power Station in Indonesia which produces 259 megawatts.

The limited locations of the most productive geothermal plants speak to the greatest weakness of geothermal energy as a stand-alone alternative energy. Not all locations on earth have the same access to geothermal resources.

In areas such as the Himalaya Mountains, where there is doubly thick continental crust of approximately 40 miles, drilling for geothermal resources would be cost prohibitive. Transmission of power from areas with geothermal resources to the many areas where they do not exist adds further costs.

On the other hand, many geologists agree that the potential exploitation of geothermal resources has barely been scratched. Among the reasons for that situation are problems such as the creation of earthquakes, the potential cooling of some geothermal resources over time, minor releases of greenhouse gases and corrosion of pipes caused by the fact that geothermal waters often carry dissolved corrosive fluids that accumulate as the waters cool.

The corrosion might be addressed by a new type of pipe made from a magnesium alloy.

Finally, and happily, one type of geothermal energy, the use of geothermal for home heating, is available worldwide. In virtually every location on earth at a depth of nine feet, the temperature is almost always 50 degrees Fahrenheit. Were every dwelling to employ the strategy of harnessing this passive geothermal energy it could address up to one half of our current energy demands.

Helium Three

Helium three, is a stable isotope of helium with two protons and one neutron. Except for Protium, Helium Three is the only isotope of any element with more protons than neutrons.

Among others the European Space Agency believes that Helium Three could be used to safely power fusion reactors. It has been estimated that as little as 17 tons of helium three might be sufficient to provide for all of the earth's power needs for a year.

Unfortunately, or rather fortunately, due to the earth's magnetic field, Helium Three, a product of the solar wind, is extremely rare on earth. That, however, is not the case with the moon.

China has been interested in Helium Three deposits on the moon at least since 2020. China's Chang'e 5 lunar mission returned from the moon with a transparent, colorless, columnar phosphate mineral containing Helium three that was subsequently named Changesite-(7).

In addition to the problematic placement of major Helium Three deposits, all of the problems that go along with the development of mineral deposits on earth are increased.

The annual shipment of 17 tons of Helium Three from the moon to the earth might be feasible in the near future with a new generation of larger rockets, Helium Three must first be located, assessed, mined and processed sufficiently to make the costs associated with its shipment feasible.

Further compounding the problems associated with lunar Helium Three, the vast majority of all of these activities would have to be conducted on the Moon. Whether lunar Helium Three exists in sufficient concentrations to make any of this possible remains an open question.

Fusion

What about fusion energy not dependent on a problematic fuel such as Helium Three? After all hasn't it been a while since the media breathlessly announced that for the first time a fusion experiment had produced more energy than it consumed? Has the breakthrough in fusion been somehow sabotaged by big oil or is the old joke that "fusion energy is 20 years away and that this will always be the case," a truism?

In fact, in December of 2022, scientists at the Lawrence Livermore National Laboratory achieved "ignition" (a reaction that released more energy than it consumed) using 192 high powered lasers to blast pellets of Deuterium and Tritium. According to the scientists the lasers caused the fuel pellets to implode and generate temperatures and pressures higher than the center of the sun. The actual energy produced was comparatively underwhelming, however, described as sufficient "to boil approximately 10 kettles of water."

In a December 22, 2022 The World Nuclear.org discussed some of the steps that would need to be achieved before Fusion energy can become a viable mass energy source.

https://world-nuclear.org/information-library/current-and-future-generation/nuclear-fusion-power

"Fusion power," World Nuclear.org writes, "offers the prospect of an almost inexhaustible source of energy for future generations, but it also presents so far unresolved energy challenges."

So, what are those challenges? Without getting to far in the weeds here, the challenge has been described as "to achieve a rate of heat emitted by a fusion plasma that exceeds the rate of energy injected into the plasma."

That problem, World Nuclear.org speculates could be achieved using "…tokamak reactors and stellarators which could confine a Deuterium-Tritium plasma magnetically."

So what is a Tokamak reactor? Wikipedia defines it as "…a device which uses a powerful magnetic field generated by external magnets to confine plasma in the shape of an axially-symmetrical torus.

Iter.org (Iter means "the way" in Latin) defines the tokamak as "…an experimental machine designed to harness the energy of fusion. Inside a tokamak the energy produced through the fusion of atoms is absorbed as heat in the walls of the vessel…"

https://www.iter.org/mach/Tokamak

The heart of a Tokamak reactor is a doughnut-shaped vacuum chamber. "Inside, under the influence of extreme heat and pressure, gaseous hydrogen fuel becomes a plasma….

"The charged particles of the plasma can be shaped and controlled by massive magnetic coils…" a means to "…confine the hot plasma away from the vessel walls."

Theoretically …"As the plasma particles become energized and collide they begin to heat up. Auxilliary heating methods help to bring the plasma to fusion temperatures (between 150 and 300 million C). Particles energized to such a degree can overcome their natural electromagnetic repulsion on collision to *fuse*, releasing huge amounts of energy."

On May 10, 2023 the MIT Technology Review reported that the Helion Energy corporation, backed by OpenAI's Sam Altman, claimed it would have a viable Fusion powered facility in operation within five years. It also reported that a number of experts doubted Altman's optimism.

Based in Everett, Washington, Helion aims to have a relatively modest 50 megawatt power plant up and running by 2028. Other Fusion start-ups are aiming for later dates in the 2030s.

https://www.technologyreview.com/2023/05/10/1072812/this-startup-says-its-first-fusion-plant-is-five-years-away-experts-doubt-it/

On February 8, 2024 the Joint European Torus (JET) laboratory in Oxford, England set a new record for fusion energy by producing 69 megajoules of energy over five seconds. The result was achieved using 0.2 milligrams of fuel and could have powered 12,000 households for that same period of time.

Landfill Gas and (other energy production that also consumers methane)

Landfill gas, the harnessing of Methane produced by the decomposition of garbage in landfills, would seem to be an ideal green energy. It not only presents an alternative form of energy production but consumes Methane, which is 28 – 90 times as potent a greenhouse gas as CO2.

The largest landfill gas power plant currently operating is at Altamont is California. The plant produces enough power for a modest 8,000 homes.

https://wmnorcalnev.com/facilities/altamont-landfill/renewable-landfill-gas-to-energy-at-altamont/

Methane, of course, not contained becomes a greenhouse gas, 28 - 90 times as powerful as carbon dioxide. Thus, the destruction of methane for power production is doubly beneficial.

Landfills, of course, are not the only sources of atmospheric methane. Unfortunately, the vast majority of those sources would not be commercially viable. Two exceptions might be commercial feed lots and sewer facilities.

Tidal Power

Currently there are no tidal power plants in the United States. Tidal plants are plants that make use of the difference or range between high and low tides to generate electricity.

According to the US Energy Information Administration (EIA), however, European tides that rose and fell up to 40 feet were used to operate grain mills more than 1,000 years ago.

Unfortunately, this form of energy is not widely available, as the commercially viable range needs to be at least 10 feet. According to EIA the only sites with such potentials are in the US are the Cook Inlet of Alaska and a number of places in Maine. Further, tidal power facilities may negatively impact tidal estuaries.

Internationally, the largest tidal power plant is the Sihwa Lake Tidal Power Station in South Korea, which produces 254 megawatts. The oldest operating tidal plant is in La Rance, France which produces 240 megawatts. Smaller plants are in Canada, China and Russia.

Solar Furnace

Solar Thermal Energy does not convert solar energy directly into electricity. It is the focusing of a significant part of the sun falling on the earth to a central point that can reach temperatures sufficient to be able to conduct high temperature metallurgy.

My father was one of the pioneers in this field. He was a believer in the legend that Archimedes destroyed the invading Roman army by having the Greek soldiers shine their shields and focus the reflected sun on the masts and sails of the Roman fleet.

A new solar thermal energy containing 30,000 moving mirrors pointed at two huge central towers is opening in Gansu province in China. Instead of solar panels the plant uses thousands of mirrors that reflect an average of 94% of the light that hits them.

The mirrors are pointed at two enormous central towers where water is turned to steam which turns turbines that generate electricity.

The solar thermal plant also contains a huge amount of molten salt, which stores heat like a battery stores electricity. When the sun is not out the heat from the salt continues to make steam.

Similar to the solar furnace is passive solar architecture, including devices such as the Trombe wall.

Thorium

Rare Earths are often discarded as radioactive waste due their frequent association with uranium and thorium. However, the thorium, at least, could potentially provide clean, safe and cheap nuclear power

Currently the EPA requires that thorium (often found in association with rare earth minerals) must be handled in a "...a very specific and costly way...".... causing rare earth mining in the United States to be "prohibitively expensive." .... however, both the environmental and economic problems associated with rare earth mining and processing in the US might be solved ...(by thorium extraction and utilization) because Thorium can be used in a special type of nuclear reactor which has been shown to be proliferation resistant and safer than the High Pressure Water Reactors which are based on Uranium.

https://www.amazon.com/Rare-Earth-Crisis-monopoly-endangers/dp/B09RV17NGC/ref=sr_1_1?crid=12K97393WG2YO&keywords=the+rare+earth+crisis&qid=1674358316&sprefix=the+rare+earth+crisis%2Caps%2C276&sr=8-1

While there are currently no operating thorium reactors, in the 1960s there was a Thorium fed Liquid Fluoride Salt Reactor at Oakridge National Laboratory that operated without incident for years until it was shut down by Congress in favor of fast breeder reactors. Some have put forward the theory that the reason for the switch from thorium to plutonium reactors had to do with the fact that, unlike plutonium reactors, thorium reactors did not produce a waste product that could be applied towards nuclear weapons production.

Cogeneration

In the original version of his book *Irrevy: An Irreverent Look at Nuclear Power*, co-inventor of plutonium John William Gofman contends that cogeneration, that is, the use of heat produced during industrial processes for electrical production, could potentially become the largest source of energy production in the world.

An example of potential cogeneration that stares anyone who lives near a refinery in the face. Generators located above and incorporating the heat of the flares that frequently burn off methane at refineries would make good use of heat that is otherwise wasted.

Coastal flooding North Carolina By North Carolina Department of Transportation - https://www.flickr.com/photos/ncdot/21972557315, CC BY 2.0, https://commons.wikimedia.org/w/index.php?curid=44019446

Conclusion: What Needs to be Done

First and foremost, the general public needs to become aware of the dire situation presented by the super-greenhouse gases and convey that knowledge to lawmakers, media, environmental organizations, other NGOs and local organizations and other individuals.

Below is a letter to the editor of a local newspaper that the reader may feel free to use for just that purpose. As most newspapers are most likely to print short letters to the editor, the below the letter is a mere 222 words.

> Editor;
>
> I have to say that I am profoundly disappointed with the COP29 UN environmental conference. The statement of the US Climate Czar acknowledged that greenhouse gases other than CO2 now accounted for approximately 50% of global warming and that an inordinate amount of effort was directed at CO2, the most benign of the the greenhouse gases.
>
> Unfortunately, the conference then went on to address only the next most benign greenhouse gases, methane (between 30 and 90 times as potent a greenhouse gas as CO2 and nitrous oxide (up to 300 times as potent as CO2) and virtually ignoring the family of super greenhouse

gases whose greenhouse effect can be many thousands of times greater than that of CO2.

For example, Nitrogen Tri-Fluoride (NF3), a gas used in the manufacturing of computers, flat screen televisions and solar panels, whose use is growing at an estimated rate of between 10 and 11% a year has a mind-bending greenhouse effect 17,200 times that of CO2, and can last in the atmosphere for 500 years. Another true super greenhouse gas Sulfur Hexafluoride (NF6) has an even greater effect, over 20,000 times that of CO2 and whose use is growing by nearly 10% per year. Shockingly, there are currently there are no international efforts to eliminate or even reduce the use of these potentially apocalyptic gases.

Sincerely

To my Senators and other political figures I have sent:

Dear Senator Bennet;

As the consequences of global warming become strikingly obvious to all who would see, it has also become obvious that our efforts to control it have failed miserably. This might well be that our efforts so far have been aimed almost exclusively at the least harmful greenhouse gas, carbon dioxide (CO2), whilst a family of super greenhouse gases, several of which are thousands of times more potent greenhouse gases are virtually ignored.

For example, nitrogen trifluoride (NF3), used in the production of computers, flat screen televisions and solar panels is 17,200 times more powerful a greenhouse gas than CO2. Worse, the use of NF3 has been increasing by at least 10% per year. A study from the University of California found that 16% of NF3 used ends up in the atmosphere where it continues to exist for at least 500 years.

It doesn't take too much imagination to see that the disaster that could follow the continuing use of NF3 would be apocalyptic.

Please ban the continuing use of NF3, and restrict imports that have used it, in their manufacture. There are alternatives. Please also investigate and prepare to ban the other super greenhouse gases, many of which are also seeing concerning growth in usage, such as sulfur hexafluoride (SF6).

Sincerely,

Richard Roy Blake
Thornton, CO 80233

If you believe a longer and more detailed letter would be most effective here is just such a letter. While it currently mentions Hurricane Helene, the most deadly hurricane to hit America since Katrina, it may be updated with more recent environmental calamities that, unfortunately, are certain to follow:

To Whom It May Concern;

> As Hurricanes Helene and Milton which both occurred in a two week window, the wildfires that have plagued Canada, Europe and the US, and the continuation of "hottest months/years," devastating floods, and droughts, demonstrate that climate change/global warming is an undeniable reality.
>
> I worry, however, that mankind's efforts to address this crisis to-date have been ill-advised, dangerous and ultimately disastrous.
>
> To date almost all of our efforts to alleviate climate change have focused on carbon dioxide, (CO_2). With the exception of water vapor, CO_2 is the most benign of the greenhouse gases. In addition to energy production, CO_2 is emitted by animals including humans, volcanoes and other natural processes. CO_2 emissions also have the advantage in that they are naturally sequestered by the many CO_2 "sinks" of the natural world, including the ocean and all plant life.

Conversely, other greenhouse gases are many times more potent global warming agents than CO_2, including methane, which is 28 – 90 times more potent, nitrous oxide, which is 300 times more potent, and most menacingly, a group of greenhouse gases known as super-greenhouse gases. One of those gases, nitrogen trifluoride (NF_3), is estimated to be 17,200 times as potent a greenhouse gas than CO_2.

Worse, the use of NF_3 is estimated to be increasing at between 10 and 11% per year. At that rate, NF_3 may overtake CO_2 as the primary greenhouse gas by 2033. It does not take an expert to realize that if global warming increases by 17,200%, human survival is unlikely.

It may well be that much of the reason global warming appears to be accelerating at a pace much greater than predicted is already attributable, to the super-greenhouse gases, in addition to increases in methane and nitrous oxide emissions emissions.

I also worry that government agencies have disseminated inaccurate information concerning the relative impacts of CO_2 and other greenhouse gases. For example; in 2022 EPA reported that nitrous oxide constituted 2,970,000,000 metric tons of CO_2 equivalent, which was a 30% increase from 1990.

Comparing nitrous oxide's CO2 equivalent with the official figures for actual CO2 releases in 2022 of 6,343,000,000, it would seem that nitrous oxide alone added an additional nearly 50% to CO2's global warming potential that year.

At the same time, however, the EPA's pie chart of greenhouse gases shows that nitrous oxide's contribution to total greenhouse gases at just 6%, whereas, it credits CO2 with nearly 80%. Yet the US Climate Czar's office pre-COP29 statement assigned only slightly more than 50% of total greenhouse gases. Clearly, something is wrong here.

Please take any steps in your power to effectively address this dire situation. For example....

(you can then suggest that the recipient of your letter take any, several or all of the below actions)

1. It seems like a no-brainer that the use of a greenhouse gas as potent as NF3 must be curtailed. While it might not be comfortable, and in fact disastrous to many, it would seem that mankind might be able to weather a climate change based on CO2 in the atmosphere alone. That would certainly seem unlikely if NF3 and

the other super-greenhouse emissions continue to grow significantly at their present rate. Previous international conventions against ozone depleting chemicals proved that the world is capable of addressing an existential threat to human survival and the environment.
2. Purchasers of solar panels, flat screen televisions etc. must be compelled to purchase these products from manufacturers that do not use any of the super-greenhouse gases, such as NF_3 in their manufacturing process.
3. A substitute for NF_3 currently exists, in the form of fluorine gas. Additionally, well-funded research might well discover, other less problematic methods for fluorine gas usage and even perhaps discover other, even less problematic substitutes. As much of NF_3 pollution occurs in east Asia, while a prohibition on its use in the West will be helpful, the more important question remains how to entice and/or force NF_3 polluters in east Asia into ceasing its use.
4. As noted, some NF_3 manufacturers insist that there is no alternative to the use of NF_3. Yet as the primary use of NF_3 is to dissolve and remove silicon, silicon dioxide, etc. defects alternatives do exist. While silicon, silicon dioxide, etc., resists many substances there are substances other than NF_3 that dissolve it. Examples

include hydrofluoric acid (HF) and strong bases such as hot potassium hydroxide and hot sodium hydroxide solutions

5. An interim step might be to require more efficient NF3 sequestration during production. This could require the use of heavy-duty fans and vacuums that direct NF3 emissions to a central point, collect them and recycle them. If production facilities are granted such a waiver their use of NF3 ought still to be drastically reduced, requiring them to become dependent on their own recycled NF3. Any such facility also ought to be frequently and effectively monitored by a number of regulatory bodies.

6. In order to sequester methane increasing amounts of methane, a greenhouse gas 28 - 90 times more potent than CO2, projects to exploit the use of gas from landfills and other methane rich environments most be much more robustly supported worldwide.

7. Nitrous oxide emissions from the use of nitrogen fertilizer could be addressed through farm policies encouraging the use of alternatives including animal manure, compost, blood meal, grass clippings, wood ash, alfalfa meal, food waste such as eggshells and banana peels, yeast brew, drip irrigation and vermicompost.

8. Well-funded and earnest efforts to explore the energy potentials of geothermal, Helium 3,

Landfill gas, Tidal power, Solar furnace, Cogeneration, Thorium and Fusion reactors. The projects must be insulated from outside lobbyists and political pressure.
9. Conduct Research into non-greenhouse alternatives for the other than NF3 "super-greenhouse gases" including, CarbonTetraFluoride, Chlorotrifluoromethane, Hexafluoroethane, Octafluropropane, Sulfur Hexafluoride, Fluoroform, Perflubutane, and any other hydrofluorocarbons, or perfluorocarbons and present legislative recommendations for their control, and the elimination of their use.

In an indication of just how urgent a comprehensive strategy to attempt to alleviate climate is necessary, the scientists at UC Irvine stepped (albeit tentatively) outside of their scientific neutrality to ask and answer the question "What can the general public do about NF3?"

A comprehensive strategy to address global warming is one that not only targets the currently most common greenhouse gases but also addresses, currently less common greenhouse gases, many of which are thousands of times more powerful than CO2.

The UC Irvine scientists answer the question thusly:

> "The public should recognize that technology has

a price and that the manufacture of these screens is not greenhouse-free. NF3 is not special; it is one of many high-tech products that potentially can have a large carbon footprint. The public should demand that the carbon footprint of such products be evaluated independently (not by the chemical manufacturers) and disclosed as energy efficiency is posted on appliances."

A comprehensive strategy for addressing global warming also needs to react to the ever-changing greenhouse gas mix. In previous chapters it was pointed out that NF3 could theoretically contribute a greater share of greenhouse gases than CO2 by 2033, and that nitrous oxide could pass CO2 by 2040.

According to volume 49 of the *International Journal of Greenhouse Gas Control,* however, a mix of gases, some of which are super-greenhouse gases have already passed nitrous oxide.

> "According to the National Oceanic and Atmospheric Administration (NOAA) (Anon, 2015b), CFC-12, CFC-11 and 15 other long-lived halogenated gases (CFC-113, CH_3CCl_3, CCl_4, HFCs 134a, 152a, 23, 143a, and 125, HCFCs 22, 141b and 142b, halons 1211, 1301 and 2402 and SF_6) account for about 0.282 W/m^2 of global RF in 2014, which is more than the global RF of nitrous oxide (0.187 W/m^2), the 3rd most

important climate forcer after carbon dioxide (1.909 W/m²) and methane (0.500 W/m²)."

Lastly, it is important to note that NF3 is not the only super greenhouse gas whose increasingly presence in the atmosphere is alarming. According to NOAA Global Monitoring, Sulfur Hexafluoride (SF6), whose potency as a greenhouse gas is the only one greater than NF3 (over 20,000 times greater than CO2), increased by 9.6% from June 2023 to June 2024.

Appendices..................................84

Appendix A

Percentages of individual greenhouse gases to the overall greenhouse gas total volume per EPA 2022 (potentially off by up to 25 to 30% per Climate Czar statement to COP29 UN Climate Conference in 2024)......................88

Appendix B

Could Super Greenhouse Gases be Used to Terraform Mars?............................89

Appendix C

Increases in Global and East Asian Nitrogen Trifluoride (NF_3) Emissions Inferred from Atmospheric Observations............................98

Appendix D

From the European Commission, Climate-friendly alternatives to HFCs...................120

Appendix E

Total U.S. CO_2 Greenhouse Gas Emissions in 2022........................126

Appendix F
UC Irvine 10/29/2008 report on NF_3.............................…...128

Appendix G

From Science Direct, Carbon Tetrafluoride, based on Chemical Engineering Journal, 2021…..………131

Appendix H

Email queries to Sierra Club, EPA, NOAA, University of California, Irvine, climate scientists, University of San Diego, climate scientists and UN Environmental Programme, responses and follow-ups..........................151

Appendix I

Fighting global warming by GHG removal: Destroying CFCs and HCFCs in solar-wind power plant hybrids producing renewable energy with no-intermittency from International Journal of Greenhouse Gas Control, Volume 49....................................185

Appendix J

Best Practices to Reduce SF6 Emissions...........................193

Appendix K

What is Nitrogen Trifluoride (NF3) Professor F. J. Martin-Torres Chaired Professor in Planetary Sciences University of Aberdeen, Scotland UK..198

Appendix L

COP29 Super Pollutants

(statement of the Climate Czar via the State Department to the COP29 UN Climate Conference 2024).............206

Appendix M

The Sprint to Cut Climate-Super-Pollutants COP29 Summit on Methane and Non-CO2 GHGS.............……..……207

Appendix N

COP28: What Was Achieved and What Happens Next?...................................219

Appendix A

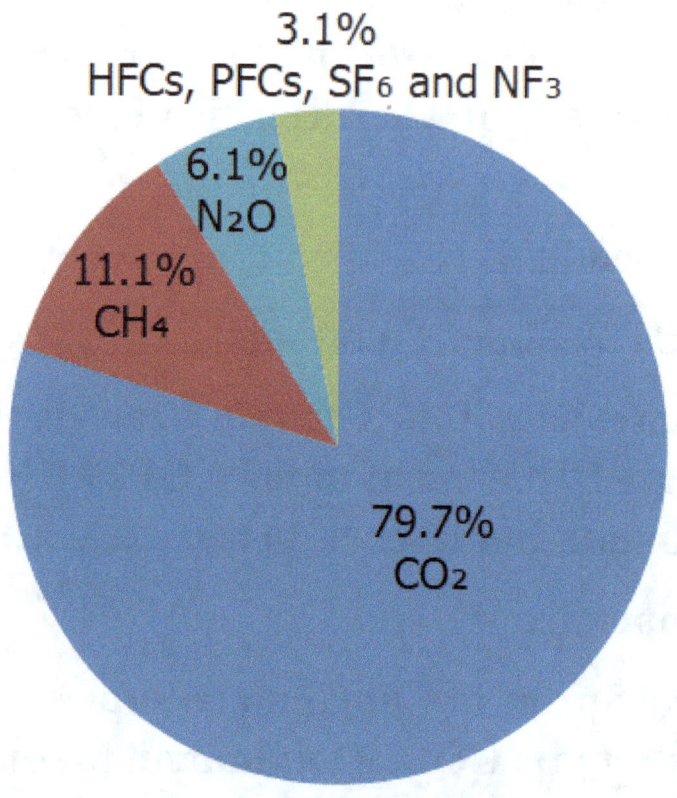

Percentages of individual greenhouse gases to the overall greenhouse gas total volume per EPA 2022 (potentially off by up to 25 to 30% per Climate Czar statement to COP29 UN Climate Conference in 2024)

Appendix B

Could Super Greenhouse Gases be Used to Terraform Mars?

In what first seems almost an attempt at gallows humor, an organization devoted to the colonization of Mars and calling itself Marspedia, suggests a massive deployment of NF3 in the Martian atmosphere to bring about a greenhouse effect on the red planet. What could go wrong? Why not ruin two planets instead of just one?

From: Marspedia

https://marspedia.org/Super_Greenhouse_Gases

Super Greenhouse Gases

Jump to: navigation, search

Super Greenhouse Gases (SGG) are hundreds or thousands of times more powerful than CO2 in warming planets, and are regulated on Earth for that reason. On Mars, which is too cold, long lived Super Greenhouse Gases (SGG) are considered an economic and desirable way to warm the planet. Types of gases which are long lived under Martian conditions are especially valuable for this purpose. [1] [2] [3] [4]

⌐

Contents

- 1 Discussion of Greenhouse Gases

- 2Super Greenhouse Gases (SGG)
- 3Compounds After the Breakdown of SGG
- 4Cost of Creating These Gases
- 5References

Discussion of Greenhouse Gases

Planetary atmospheres warm planets by allowing light to hit the world, but slows the radiation of infrared (heat energy) leaving the world. (This is known as the Greenhouse effect.) Without our atmosphere, Earth would have a sub freezing temperature of -10 C. However, not all gases warm planets equally. Some such as oxygen (O2), and nitrogen (N2) are transparent to heat energy. More complex molecules tend to slow the radiation of heat to space. Carbon dioxide (CO2) is a Greenhouse gas which is causing the Earth to warm, as it concentration increases in Earth's atmosphere. The strength of other green house gases are measured relative to carbon dioxide. For example: methane is 80 times more powerful than CO2 during the 20 years it is expected to remain in the atmosphere. Water (H2O) is a powerful greenhouse gas, but it rapidly leaves the atmosphere as rain and snow. Carbon dioxide remains in the air for a long time. (CO2 is expected to last in the air for about 200 years, when it is typically absorbed by some sort of plant. However, the CO2 is returned to the air when the life rots a few years later. To draw down the CO2 permanently, it needs to be removed from the atmosphere AND the biosphere.) If we introduce life onto Mars, methane (CH4) would be added to the atmosphere which also is a greenhouse gas.

Many gases will help retain heat, but each is best at trapping specific wavelengths of infrared radiation. To warm Mars, we would wish to pick gases which block 'windows' in the spectrum where heat can escape from Mars. (There is plenty of CO2 on Mars, and if it warms, then water (H2O) will also be more common in the air. So gases which block wavelengths that these two compounds don't, are of special interest.

Super Greenhouse Gases (SGG)

If kilotonnes of gases are to be created then we want to create compounds which will be stable in the Martian atmosphere for many decades or centuries. Mars is bombarded by Ultraviolet light the most energetic of which has energies sufficient to break molecular bonds. Fluorine (F) has the strongest chemical bonds, so its compounds are ideally suited for SGG in Mars' atmosphere. Unfortunately, fluorine is fairly rare. Chlorine (Cl) is chemically similar and much more common and cheaper, so some fluorine atoms may be replaced with Cl, as a cheaper, less long lived substitute. However, Chlorine can destroy the ozone layer, so in the long term it should be avoided. Several industrial refrigerants such as 1,1,1-Trichloroethane, have been suggested, but with Cl as part of the structure, they will fight the goal of building up Mars' Ozone layer.)

Because these gases absorb different parts of the infrared spectrum, it is likely that the ideal solution for warming Mars would be a mix of several chemicals. (Mixtures of CF_4, C_2F_6, C_3F_8, and SF_6, are the most common gases looked at in the literature.)

Large molecules tend to have better warming potential (they can hold heat by more vibration modes between the atomic bonds), but are more likely to be broken up by Ultraviolet light, so they last for shorter periods in the atmosphere. Unless a short term 'burst' of warming is wanted, longer lived molecules will likely be preferred.

Fogg, in his book on Terraforming, points out that these gases are most useful at low concentrations. (At higher concentrations, you reach a region of diminishing returns.) This suggests that SGG would be used as a first step in terraforming, useful in the initial goal of destabilizing the South Pole CO_2 ice cap.

Climate modelling suggests that the warming would be strongest near the equator and in low lands. If we got a 20K increase in temperature planet wide, at the bottom of the Hellas basin would be about double that. (A five degree increase at the South Pole would destabilize the southern carbon dioxide ice cap. Note that the South Pole is 1 to 3 km above the average height of the planet, and is far from the equator, so warming will be lower there. I've not been able to find what the MINIMUM average temperature would destabilize the antarctic CO_2, but in several papers authors have said that an average 20K increase would be more than enough.)

The following table shows several chemicals, their greenhouse gas warming potential (relative to CO_2), and their expected lifetime in the Martian air. (Relative warming looks at how much stronger the greenhouse gas is compared to CO2, over a 100 year period.) The lifespan is for Earth, there the lower atmosphere is protected by the Ozone layer. Mars gets 43% as much sunlight as Earth (so gases will last longer), but lacks an ozone layer (so the gases will break up more quickly). If life on Mars develops an ozone layer, the lifespan there will be higher than on Earth, for now it is likely to be lower.

Name	Formula	Relative Warming	Lifespan (Years)	Notes
Carbon Tetrafluoride	CF4	5,700 (times more than CO2)	50,000 y	Also known as Tetrafluoromethane, or R-14
Hexafluoroethane	C2F6	11,900 or 9,500?	10,000 y	Etchant in semiconductors. Also a refrigerant when mixed with trifluoromethane.
Octafluoropropane	C3F8	8,600 or 24,000?	2,600 y	Used in semiconductor production. Other than lower life span, an ideal gas.
Octafluorocyclobutane	C4F8	10,000	3,200 y	Ring of 4 carbon atoms with Fl everywhere else. Semiconductors,

				refrigerant.
Perflubutane	C4F10	8,600	2,600 y	Fire extinguishers, and used in ultrasound contrast agents.
Sulfur Hexafluoride	SF6	22,200	3,200 y	Used in transformers and Magnesium production
Chlorotrifluoromethane	CF3Cl	14,000? or 10,800?	640 y	Chlorine can destroy ozone layer
Fluoroform	CHF3	12,000	260 y	Used in semiconductor industry and as a refrigerant
Nitrogen Trifluoride	NF3	17,200? or 12,300?	740 or 550 years?	Semiconductors, toxic in high concentrations (~1,000 ppm).

Note: different references have very different values for the relative warming. First value is from IPCC report (100 year warming potential), second value is the one most commonly found. Trying to resolve why some of these values are so different.

The very long lifespan of carbon tetrafluoride (CF4) and its high relative warming makes it attractive.

The paper "Keeping Mars Warm with New Super Greenhouse Gasses":

https://www.pnas.org/content/98/5/2154

.. looked at a number of chemicals, and studied the following 5 gases as the most likely useful:

- SF4(CF3)2
- CR3CF2CF3
- CF3SCF2CF3
- CF5CF3
- SF6

These were picked since they covered a lower set of frequencies than the carbon - fluoride compounds listed above. However the large complex molecules broke down more quickly, and were intended to be used after Mars had an ozone layer to protect them.

After studying these gases they suggested that sulphur hexafluoride (SF6) was the best to use of the ones studied.

This paper: Radiative-convective model of warming Mars with artificial greenhouse gases [5]

.. suggested that C3F8 had an absorption spectrum which was well suited to the lower pressure atmosphere of Mars. (They said on Mars, it had a higher relative warming potential, higher than SF6.) Given that this would be an ideal gas, other gases were compared to it under Martian conditions and it was found that:

-- CF4 was 17% as good as C3F8,

-- C2F6 was 49% as good, and

-- SF6 was 48% as good as C3F8.

An ideal mixture of these 3 gases with C3F8 was 16% more effective at warming the planet than C3F8 alone.

They concluded that their energy balance calculations suggest that the addition of ~0.2 Pa of the best greenhouse gases mixture, or

~0.4 Pa of C3F8 alone, would shift the equilibrium by 20 degrees K. This would be enough that CO2 would no longer be stable at the Martian poles and a runaway greenhouse effect would result. (Another study I read suggested ~1 Pa would be needed for this large of an increase in temperature, but they used a different mix of gases.)

Compounds After the Breakdown of SGG

Carbon tetrafluoride (CF4) is an ideal greenhouse gas, but eventually it will be broken down by Ultraviolet (UV) light by losing a Fluorine atom. The fluorine is HIGHLY reactive, likely combining with CO2 to form COF (which still missing a valence electron, so it will perhaps collect an H, or an OH eventually). The CF3 will eventually form another molecule such as CF3OH or CHF3.

These 'frangment' molecules will also be greenhouse gases, but little is know about them as far as their lifetime in the Martian atmosphere, or their greenhouse potential. This should be a subject of future study.

Cost of Creating These Gases

As Martian industry expands, these gases will be created for various purposes. For example, SF6 will be created for use in electrical transformers. On Earth, great care (and expense) must be taken to prevent it from leaking into the air. On Mars such effort is not needed since we WANT the planet to warm. Thus small amounts of these chemicals will inevitably build up, for 'free', in the Martian atmosphere as Mars' industry develops.

That said, here are the 2021 prices for some of the compounds listed above:

- CF4.. $7,000 / tonne. This chemical is very easy to produce, (just burn carbon in fluorine). The high price likely reflects no widespread industrial use.
- C2F6.. $150 / tonne
- C3F8.. $500 to $700 / tonne
- SF6.. $6,000 / tonne
- CHF3.. $45,000 / tonne

(Note, these prices are from quick google searches. Someone more knowledgable please correct these values if they are off.)

If industries developed to mass produce the gases by the kilotonne, these prices would drop.

The 'Keeping Mars Warm' paper discussed above suggests that no more than 170 kilotonne per Earth year would be needed to make up for the loss of these gases by UV photolysis. If we assume the average cost of these gases is $1,000 / tonne, and we want to produce 500 kilotonnes / year to build up the concentration, then the price to engineer Mars' atmosphere would be $500 million per year. This price would drop to a maintenance level of about 1/3 that, once we reach the target temperature.

References

"Terraforming: Engineering Planetary Environments", by Martin J. Fogg, ISBN 1-56091-609-5.

Wanke, H. & Dreibus, G. (1988) Phil. Trans. R. Soc. London A 235, 545–557. // Paper suggests fluorine may be more common on Mars than on Earth.

// Paper discussing which SGG would be idea for terraforming Mars.

https://www.pnas.org/content/98/5/2154

M. Gerstell, J. Francisco, Y. Yung, C. Boxe, and E. Aaltonee. Keeping mars warm with new super greenhouse gases. Proceedings of the National Academy of Sciences, 98(5):2154–2157

Y. L. Yung and W. B. DeMore. Photochemistry of planetary atmospheres. Oxford University Press, 1999. ISBN 9780195105018.
URL http://books.google.com.au/books?id=Q4pHLv9TvksC.

1. ↑ https://www.esa.int/gsp/ACT/doc/ESS/ACT-RPR-ESS-2013-IAC-MarsClimateEngineering.pdf
2. ↑ https://www.pnas.org/doi/10.1073/pnas.051511598
3. ↑ https://ui.adsabs.harvard.edu/abs/2000JBIS...53..235M/abstract

4. ↑ https://agupubs.onlinelibrary.wiley.com/doi/full/10.1029/2004JE002306
5. ↑ https://www.researchgate.net/publication/251438006_Radiative-convective_model_of_warming_Mars_with_artificial_greenhouse_gases

- This page was last edited on 23 July 2022, at 09:37.

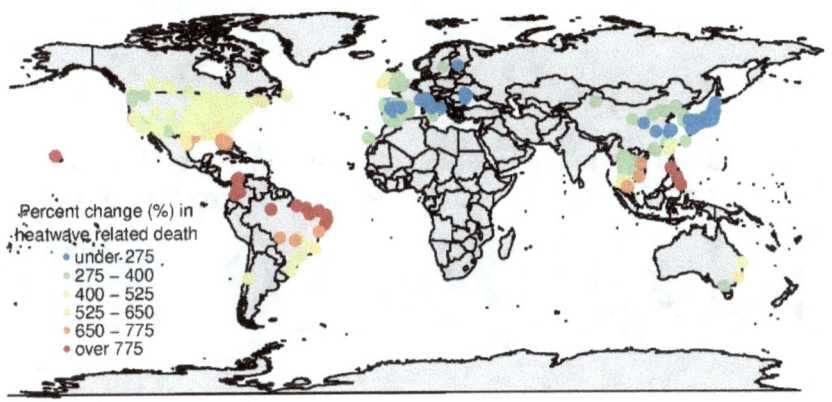

Map showing percentage change in heatwave related deaths 2022

Appendix C

Increases in Global and East Asian Nitrogen Trifluoride (NF$_3$) Emissions Inferred from Atmospheric Observations

Synopsis

Global and regional NF$_3$ emission have increased since 2015, and the majority of these emissions are from East Asia.

Introduction

Nitrogen trifluoride (NF$_3$) is a very potent long-lived greenhouse gas (GHG), with an atmospheric lifetime of approximately 569 years (1) (uncertainty in the lifetime is thought to be large (2,3)), and an extremely high global warming potential (17,200 on a 100-year time scale). (4,5) It is primarily emitted during NF$_3$ production and from its end use in the cleaning of silicon-containing deposits in the semiconductor industry or during the manufacture of flat-panel displays (FPD), such as liquid crystal displays (LCDs) and amorphous-Si/crystalline-Si thin-film photovoltaic (PV) cells. (6–9) Minor emissions are thought to be associated with its use as a rocket fuel oxidizer, as a fluorine donor for chemical lasers, and for fluorochemical production. (6,10) Leakage during its transportation likely has a negligible contribution to the atmospheric NF$_3$ abundance. (8) The removal of NF$_3$ in the atmosphere is mainly through photolysis in the stratosphere and mesosphere. (11,12)

NF$_3$ was included in the basket of substances controlled under the Kyoto Protocol through the Doha Amendment in 2012. Some countries have specific targets to reduce NF$_3$ emissions as part of their wider greenhouse gas emissions reduction strategy, e.g., Japan aimed for reduction from 1.6 million tons CO$_2$-eq in 2013 to 0.5 million tons CO$_2$-eq in 2030. (13) NF$_3$ emissions have also been required to be reported to the Chinese government since 2021. (14) South Korea, one of the major emitters, (7) does not report its NF$_3$ emissions to the United Nations Framework Convention on Climate Change (UNFCCC) or include it in their commitments to the Paris agreement. (15) Other Annex I Parties (such as USA, Canada, and countries in the European Union and Oceania) have a bulk reduction commitment to UNFCCC, but do not single out NF$_3$ emissions reduction. (16−20)

Previous studies showed that global NF$_3$ emissions had risen from undetectable levels in the 1980s to 1.18 ± 0.21 Gg·yr^{-1} in 2011, based on inverse modeling using a 12-box atmospheric chemistry transport model combined with atmospheric measurements. (6) The latest World Meteorological Organization Scientific Assessment of Ozone Depletion reported that NF$_3$ emissions increased from 2.0 ± 0.1 Gg·yr^{-1} in 2016 to 3.0 ± 0.1 Gg yr^{-1} in 2020, based on results from the 12-box model and measurements at five AGAGE stations. (5)

Emissions from East Asia were estimated for 2014 and 2015 by Arnold et al. (7) In that study, emissions from South Korea were estimated to be approximately 0.4 ± 0.05 and 0.6 ± 0.07 Gg yr^{-1} in 2014 and 2015, respectively. Emissions from China were found to be potentially substantial but were inferred with very large uncertainties (1.08 ± 1.17 and 0.36 ± 1.36 Gg yr^{-1} in 2014 and 2015, respectively).

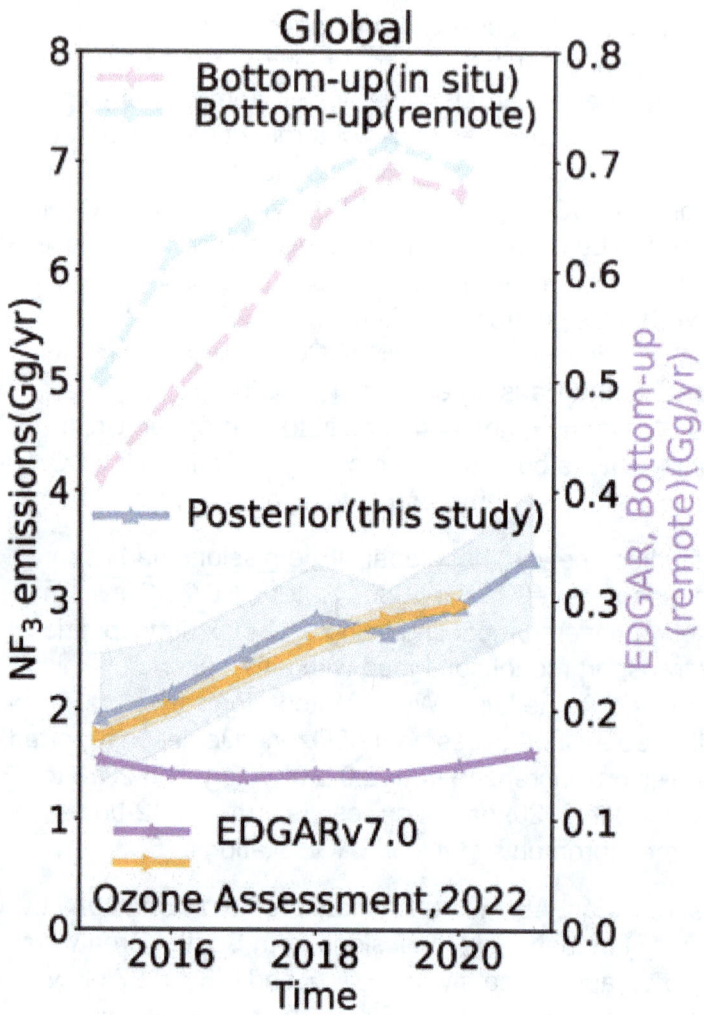

Growth in Global Mole Fractions

A previous study reported that the global tropospheric annual mean dry air mole fraction increased from almost zero in the early 1970s (0.008 ppt in 1975) to 0.86 ± 0.04 parts per trillion (ppt) in 2011, based on archived air data samples. (6) Updated reported values, including in situ measurements, show that this growth continued, reaching to 2.3 ppt in 2020. (5) Data updated here from observations at AGAGE measurement sites show that NF_3 global mean mole fractions reached 2.59 ± 0.11 ppt in 2021 (Figure 1). The mean growth rate of the global mole fraction between 2015 and 2021 was 0.22 ± 0.12 ppt yr^{-1}. This value reached a maximum in 2021 (0.28 ± 0.15 ppt yr^{-1}), suggesting that global emissions have continued to grow throughout this period.

Figure 1. Monthly mean NF_3 mole fractions and their standard deviations in monthly variability measured at (a) seven AGAGE stations in relatively remote locations: CGO: Cape Grim, Tasmania; SMO: Cape Matatula, American Samoa; RPB: Ragged Point, Barbados; THD: Trinidad Head, California; JFJ: Jungfraujoch, Switzerland; MHD: Mace Head, Ireland; ZEP: Zeppelin, Norway; (b) monthly mean NF_3 mole fractions and their standard deviations in monthly variability measured at GSN: Gosan, South Korea; (c) global annual mole fraction and annual growth rate. Global annual mole fractions are from the average of the monthly mean NF_3 mole fractions from five background sites (CGO, SMO, RPB, THD, MHD).

Increases in Global Emissions

Here, we report inferred emissions from three inversions using three different a priori spatial distributions. The prior annual mean emissions are extrapolated from Arnold et al. (6) The three prior distributions are derived from proportions of semiconductor (wafer) capacity, flat screen display capacity (as most of the flat screen display production is for liquid-crystal displays (LCDs), we assume that all flat screen production is for LCD throughout), or NF_3 market share for each region (we assume NF_3 market share represents the NF_3 consumption from each region in Supplementary Table 2 and regions are defined in Supplementary Figure 1). Global posterior emissions are the total of the derived NF_3 emissions from 11 regions defined in Supplementary Figure 1. The amount of NF_3 employed in the photovoltaic (PV) cell industry only accounts for a small fraction of global NF_3 production (about 3% (8)), so we do not include this sector in our prior emission distribution estimates.

The mean inversion results show that global emissions of NF_3 rose from 1.93 ± 0.58 Gg yr^{-1} in 2015 to 3.38 ± 0.61 Gg yr^{-1} in 2021, with an average annual increase of 10% yr^{-1}, with large error reduction (>80%) compared to the prior uncertainty (Figure 2 and Supplementary Figures 2–5). Note that the average of the inversion results using the three different prior estimates is presented, unless otherwise stated. The overall emission growth rate derived using a linear regression for the time period 2015–2021 was significantly positive, at 0.70 ± 0.06 Gg yr^{-2}.

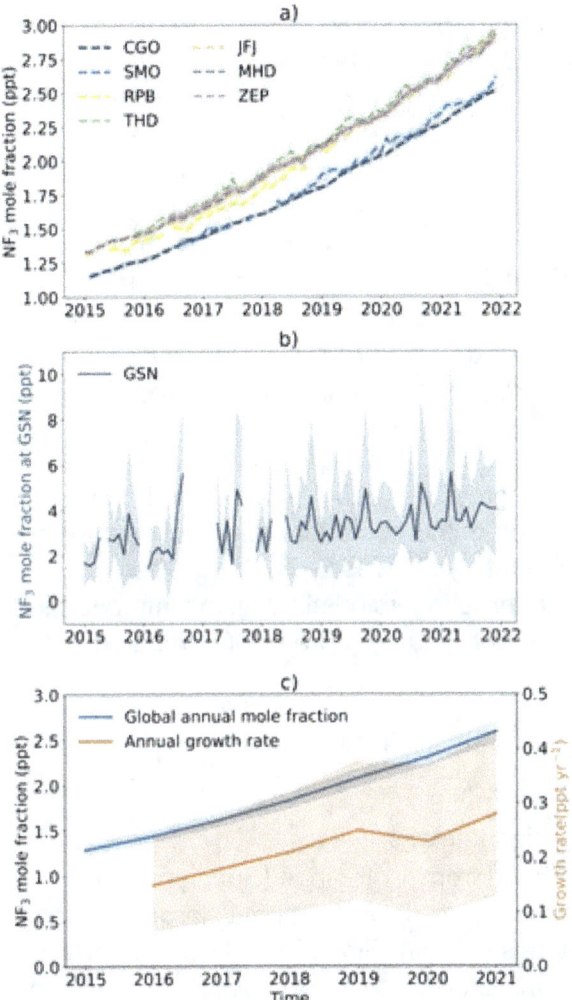

Figure 2. Comparisons of top-down from this study to bottom-up estimates and previous studies: (a) Global compared to EDGAR v7.0 (26) and Laube & Tegtmeier (5); (b) China (with Taiwan included) compared to EDGAR v7.0, Guo et al. (27) and Arnold et al. (7); (c) South Korea compared to Arnold et al. (7); (d) Japan compared to data from EDGAR v7.0, Japan's 2023 National Inventory Report to the UNFCCC (NIR, Japan), (10) Arnold et al. (7) For (a)

Global and (d) Japan, we also show bottom-up (in situ) (in situ meaning low NF_3 conversion rate) and bottom-up (remote) (remote meaning high NF_3 conversion rate) emissions estimated in this study with no abatement (destruction of NF_3 in exhaust gas) considered, while NIR (Japan, 2023) data state that abatement was considered. Top-down results and bottom-up (in situ) are sharing the left y-axis, while EDGAR v7.0, NIR, bottom-up (remote) are using the right y-axis. Note that the left y-axis is 10 times higher than the right y-axis and bottom-up (in situ) results are around 10 time larger than results from bottom-up (remote).

The model simulations agree well with the observed trend, as shown by the comparison of the posterior mole fraction and the observations (Supplementary Figures 6–7). Supplementary Table 3 shows that the three inversion simulations improve the model and observation root-mean-square error compared to the prior.

Emissions from East Asian Regions

Emissions from East Asian Regions
Emissions were estimated from 11 regions (Supplementary Table 4). Among these regions, East Asia emits large amounts of NF_3 to the global totals (>70%). Since East Asia is the main contributor to global emissions and their variability, this section focuses on that region.

We separately quantify East Asian emissions from south China, northeast China, South Korea, and Japan. Emissions from the rest of China and North Korea are negligible, and these two regions are included into the region defined as the rest of Asia, Russia, and middle east (Figure 2, Supplementary Figures 2–3). The inferred mean annual emissions from south China and northeast China are 0.83 ± 0.13 and 0.45 ± 0.12 Gg yr^{-1} between 2015 and 2021, respectively. The contribution of the mean annual emissions from these two regions to the global emissions (2.64 ± 0.51 Gg yr^{-1}) varies between 22% and 45% for south China and 11% and 27% for northeast China over this seven-year period. South Korea has an average emission of 0.48 ± 0.08 Gg yr^{-1} between 2015 and 2021, and its emissions comprise 12–26% of the inferred average global mean emissions over the seven years. Given the proportion of semiconductors and flat panel display-LCDs manufactured in South Korea relative to global totals (24% for LCD capacity in 2018, 26% for semiconductor capacity in 2015), (22,23) we conclude that this ratio is reasonably consistent with the industrial activities in South Korea. The emissions from Japan are lowest in this region, with average emissions of 0.32 ± 0.10 Gg yr^{-1}, contributing 7–20% of global emissions.

NF$_3$ emissions have increased between 2015 and 2021 by 0.34 ± 0.18 Gg yr^{-1} and 0.26 ± 0.17 Gg yr^{-1} in south China and northeast China, respectively, while global emissions have risen by 1.45 ± 0.84 Gg yr^{-1} (see Supplementary Table 5). Hence, NF$_3$ emissions growth from China contributed around 41% of the global rise. Emissions growth in South Korea (0.27 ± 0.12 Gg yr^{-1}) and Japan (0.19 ± 0.16 Gg yr^{-1}) over this period contributed 19% and 13% to the rise, respectively. The increases from these three countries together have contributed approximately 73% of the global increase (Supplementary Table 5).

The influence of COVID-19 on the estimated NF$_3$ emissions seems small in South Korea; even during the 2020–2021 period in which there were national lockdowns, and NF$_3$ related activities, such as factory production, trading, were less active, the emissions still increased. The derived NF$_3$ emissions between 2020 and 2021 increased by 0.15 ± 0.12 Gg yr^{-1} compared to the period of 2015 to 2018 (2019 is not considered here as there were trade conflicts between Japan and South Korea in 2019, (24) when the limited raw material supply to produce semiconductors might affect the NF$_3$ consumption related activities in South Korea). In contrast, we do not find a significant rise in emissions from South China, Northeast China, and Japan during the COVID-19 period (2020–2021) compared to the period before (2015–2019) (0.12 ± 0.19 Gg yr^{-1} for South China, 0.07 ± 0.17 Gg yr^{-1} for Northeast China, 0.03 ± 0.16 Gg yr^{-1} for Japan).

Regions outside of East Asia contributed around 21% to the global averaged NF$_3$ emissions and 27% to the global emission increase (Supplementary Figures 2–3).

Among the rest of the regions, only three regions have significant emission growth between 2015 and 2021: 13% (0.19 ± 0.16 Gg yr^{-1}) from North America, 12% (0.18 ± 0.09 Gg yr^{-1}) from Europe, and 4% (0.06 ± 0.4 Gg yr^{-1}) from Oceania. Only North America released above zero emissions averaged over 7 years (2015–2021) (0.29 ± 0.10 Gg yr^{-1}). The estimated averaged emissions in Europe and Oceania are not significantly different from zero.

Southeast Asia emitted 0.18 ± 0.13 Gg yr^{-1} on average, with no significant increase in the estimation in 2021 compared to 2015. The estimated annual mean emissions decreased in between 2020 and 2021 compared to 2018 and 2019, which might reflect the influence of lockdown on the NF$_3$ related industrial activity in this region during the COVID-19 period.

The remaining regions (South Asia, Africa, Central and South America) neither have large emissions, nor significant emissions growth. The emissions from South Asia (defined as India, Sri Lanka, Pakistan and Bangladesh) have changed from 0.15 ± 0.10 Gg yr^{-1} in 2015 to 0.08 ± 0.13 Gg yr^{-1} in 2021, with emissions reduced by around 5%, which is offset by the increase in Southeast Asia. The mean emission for Africa is 0.02 ± 0.11 Gg yr^{-1} over 2015 and 2021, and the average emissions in Central and South America is not significantly from zero. This is consistent with the low NF$_3$-related industrial activities in these two regions over this time period (see Supplementary Table 2 and 4). Supplementary Figure 6 shows the error reduction in each region. Even though there are no AGAGE continuous observation network in these regions (South Asia, Africa, Central and South America, Southeast Asia), the errors are still reduced based on constraint from other observations.

DISCUSSION

Global emissions derived over the period 2015–2020 using five AGAGE stations and the AGAGE 12-box model are consistent, within 1-sigma uncertainties, with our global estimates (Figure 2). (5) The two top-down methods suggest that emissions of NF_3 are more than 10 times higher than the EDGAR v7.0 estimate during this period (Figure 2) (EDGAR, 2022). Moreover, top-down emissions have continuously increased since 2015, unlike those from EDGAR v7.0. The discrepancy between top-down and EDGAR v7.0 estimates might be due to underestimation of emissions factors by EDGAR v7.0 or emissions from certain source sectors or countries might be missing. For example, EDGAR v7.0 does not include any NF_3 emissions in South Korea, even though this country plays an important role in the semiconductor and LCD industries; EDGAR v7.0 only considers NF_3 emissions from the electronic sectors, while other processes, such as the process of producing NF_3, and other industry activities that use NF_3 (e.g., NF_3 usage in lasers), which may significantly contribute to the NF_3 emissions, are not included in EDGAR v7.0. (6,10,30)

The derived NF_3 emission for China (with Taiwan included) in this study in 2015 is statistically consistent with that was estimated in a previous study (0.98 ± 0.15 Gg·yr^{-1} compared to 0.36 ± 1.38 Gg·yr^{-1}, as shown in Figure 2 and Arnold et al. (7)), although the previous study is very uncertain. Our mean posterior estimate for China is more than twice that from Arnold et al. (7) Our estimate for South Korea (0.35 ± 0.07 Gg·yr^{-1}) is significantly smaller than that of Arnold et al. (7) in 2015 (0.6 ± 0.07 Gg·yr^{-1}), while the estimate for Japan (0.20 ± 0.1 Gg·yr^{-1}) is comparable to their top-down result (0.11 ± 0.39 Gg·yr^{-1}). Similar to our significantly higher global NF_3 top-down emissions compared to EDGAR v7.0, the regional top-down estimated emissions for China and Japan are an order of magnitude higher than the emissions estimated by EDGAR v7.0 (Figure 2). The annual mean estimates in this top-down study suggest that emissions from Japan declined during 2017–2018, consistent with the trend estimated by EDGAR v7.0 and the National Inventory Report to the UNFCCC (NIR, Japan), although higher absolute emissions for Japan are derived in this study (NIR estimate of emission reductions between this time frame were mainly driven by the reduction in production emissions, shown in the method section Bottom-Up Estimates for Japan). (10,26)

Robustness of Inversion Results

Sensitivity tests were performed to assess whether inversion results are sensitive to the network configuration, prior emissions magnitude, trend and uncertainty, and observational error (Supplementary Figures 8–13). The only significant difference from our main results is in the emissions from South Korea when observations from GSN station are not included. As there are some data gaps at this station in 2016, 2017, and 2018 due to instrumental damage by typhoons, (25) we do not know whether the variability in NF_3 emission from South Korea is caused by the data gaps or by actual changes in emissions.

It is also possible that emissions from small countries such as South Korea may be impacted by model resolution. It is difficult to assess the impact of model resolution alone, but the sensitivity of South Korean emissions to the inclusion or exclusion of Gosan, and the difference of our results with Arnold et al. (2018) may in part be related to the relatively course resolution of the model compared to the size of that country and its proximity to the Gosan.

Relating Emissions to Industrial Activity

Our global estimated posterior NF_3 emissions account for less than 10% of the global NF_3 production ([Figure 3](#) upper panel, [Supplementary Table 6](#)), and this emission ratio, defined as posterior global NF_3 emissions divided by global production, was used to explain the change of the efficiency of industrial processes. [(6)](#) The emission ratio has significantly declined between 2017 (0.09 ± 0.02) and 2020 (0.07 ± 0.01). The emission ratio in 2019 is also significantly smaller than the ratio for 2011, as calculated by Arnold et al. [(6)](#) This sustained reduction of the emission ratios could point to lower NF_3 production related emissions, higher NF_3 use efficiencies and/or more widespread of remote plasma sources (RPS) or more efficient abatement measures in the waste streams of facilities that use NF_3. [(31−33)](#)

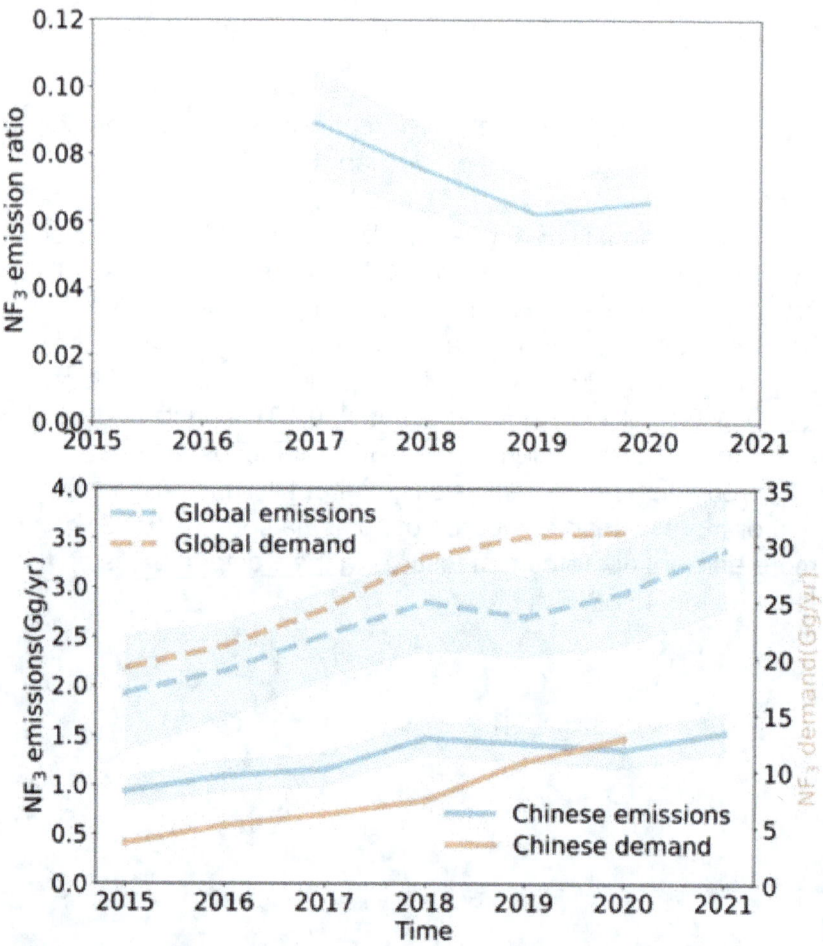

Figure 3. Upper panel: emission ratio (unitless) of posterior global NF_3 emissions to global production. Global production is the total annual amount of NF_3 produced globally. Lower panel: comparison of global and Chinese NF_3 emissions and demand. Demand is the total amount of NF_3 needed or required by the industrial companies globally or in China. The global NF_3 production data, global demand data between 2017 and 2020 and Chinese demand (not including Taiwan) are from m.huaon.com, [28] and the global demand data in 2015 and 2016 are from Adams. [29]

The estimated global posterior emissions follow a similar trend to global NF_3-related industrial activities, i.e., NF_3 demand data (Figure 3 lower panel). The significant increase in Chinese emissions until 2018 is consistent with increasing Chinese NF_3 demand data (Figure 3 lower panel, Supplementary Table 7). Our estimated Chinese emissions do not increase as fast as Chinese NF_3 demand after 2018, which might point to lower NF_3 production-related emissions, higher NF_3 utilization efficiency, or more widespread or more efficient abatement.

We derived two bottom-up emission estimates for Japan based on the NF_3 production and consumption data in the Japanese NIR. In both estimates, we consider zero abatement to assess the upper limit of emissions but apply different usage rates based on two NF_3 consumption methods: (a) in situ, in which the NF_3 cleaning process is analogous to the process for other cleaning gases like C_2F_6 and C_3F_8, characterized by low NF_3 use efficiency and high NF_3 emissions without abatement); (b) remote, in which remote plasma clean is used, characterized by high NF_3 use efficiency and low NF_3 emissions without abatement. If we assume that 100% of the NF_3 is consumed by the semiconductor and LCD production in the in situ method, our estimated upper limit emissions would have increased from 0.463 to 0.680 Gg·yr^{-1} from 2015 to 2021 (Figure 2 d). In most years, these values would be higher than the top-down emissions. If we assume that 100% of the NF_3 is used in the remote method, our estimated emissions would have only increased from 0.067 to 0.069 Gg·yr^{-1}, which would be about 80% lower than our top-down emissions estimate, but still higher than the NF_3 emissions reported in the Japanese NIR. The Japanese NIR does not report the proportion in which the two methods are used, but since our top-down emission estimate is significantly higher than the Japanese NIR emissions estimate, we suggest that some of the assumptions in the Japanese NIR, such as abatement rates, and/or the data used in the NIR, such as estimated emissions from production and industrial activities, need to be revised. Estimated emissions using the in situ and remote method from Japan account for around 10% and 9% of our bottom-up estimated global emissions using similar parameters (details shown in the Bottom-up Estimates section in Methods), respectively, which are consistent with our top-down estimated ratio (7–20%, as shown above). Our results suggest that real-world abatement efficiencies may be lower than that used for estimating Japanese NIR emissions, in line with the conclusions derived for CF_4 and C_2F_6 by Kim et al., (34) even though NF_3 is much easier to abate than CF_4 or C_2F_6.

Our derived global emissions increased from 1.93 ± 0.58 Gg yr^{-1} ($\pm$one standard deviation) in 2015 to 3.38 ± 0.61 Gg yr^{-1} in 2021, significantly larger than that from a bottom-up inventory (EDGAR v7.0). The derived emissions are also significantly higher than EDGAR v7.0 estimates in China and Japan. East Asia contributes to around 73% of the global NF_3 emission increase from 2015 to 2021, and North America, Oceania, and Europe contributed most of the remaining growth. We could only estimate the bottom-up emissions for Japan and evaluate Japanese NIR emissions in this study, as most countries do not report their NF_3 consumption and production data. We find that the uncertainties for the bottom-up estimates are difficult to calculate due to the incompleteness of the data. It would be very helpful if details such as NF_3 consumption and production data, usage rates, abatement rates, etc. were included in future NIRs. When combined with top-down emission estimates such as those presented here, this data would provide new insights to improve bottom-up methods and resulting emission inventories.

METHODS

Measurement Data

The Medusa AGAGE gas chromatograph with mass spectrometric detection (GC-MS "Medusa") was the first instrument used to regularly monitor in situ NF_3 dry mole fractions worldwide. (21,35) Continuous atmospheric

NF_3 measurements were implemented at AGAGE sites since 2011, with ambient measurements roughly every 2 h and precisions that improved from ~2% to ~1% (or better) due to improvements in analytical methods, chromatographic peak sizes, and mass spectrometer technology. (21,35) NF_3 measurements are reported on the Scripps Institution of Oceanography (SIO) SIO-12 primary calibration scale. (21) Every ambient air measurement is followed by a working standard measurement to account for changes in detector response. Each working standard is compared with a tertiary standard four times a week. Calibrated tertiary standards are sent by SIO to each measurement site and they are reanalyzed at SIO upon their return. Information for the measurement sites used for the inversion is listed in Supplementary Table 1. Mole fractions at the eight AGAGE stations are used to estimate global and regional emissions, while unpolluted filtered data are used to generate the global initial conditions. The unpolluted filtered data for a particular day are identified as the data within median plus 2σ, after removing seasonal variation using a moving window with 121 days periods (60 days before and 60 days after the current day). (36) Global annual mean mole fractions are calculated based on global monthly nonpolluted mean mole fractions and were taken from the AGAGE Web site https://agage.mit.edu/.

Atmospheric Transport Modeling Using GEOS-Chem

GEOS-Chem is a 3-D global atmospheric chemistry transport model, and we use version 12.5 in this study to conduct the forward simulation and generate the sensitivity matrix used in the inversion. (37) The forcing meteorological data are from Modern-Era Retrospective Analysis for Research and Applications, Version 2 (MERRA2) generated by the NASA Global Modeling and Assimilation Office (GMAO), (38) which drives the model with a resolution of 2.5° longitude and 2° latitude and 45 vertical levels (Supplementary Figure 1). To initialize our simulation, we spun up the model for 5 years prior to the study period, using the prior emissions extrapolated from

Arnold et al. (6) At the end of the spin-up period, the simulated mole fraction field was adjusted so that the average mole fraction at the measurement station locations was consistent with the observations. Emissions derived in the first year (2014) of the inversion were discarded to account for any residual impacts of model spin-up.

Offline fields that describe the stratospheric loss of NF_3 were not readily available for use in this study. However, given its very long lifetime compared to the study period, it is unlikely that errors in the stratospheric loss distribution will have a substantial impact on our results. Therefore, for convenience, we decided to simply rescale the stratospheric loss for some readily available species (CH_4 in this case), to provide a reasonable lifetime for NF_3. We iteratively adjusted the loss field until a lifetime of around 531–593 years was achieved (see Supplementary Figure 15). Tests using a substantially larger (around 800–1000 years) lifetime does not result in significant changes in derived emissions (differences less than 0.1% in the global total).

A Priori Emissions and State Vectors

The a priori global annual emissions from 2014 to 2022 are extrapolated using the regressed trend from the results for 2007–2011 from Arnold et al. (6) The total NF_3 emissions in each year are assigned to 11 different regions based on the prior distribution information from three different resources to generate three inversions: proportions for wafer capacity, flat screen display (LCDs) capacity and NF_3 market share. Information on the NF_3 proportions in each region is summarized in Supplementary Table 2. Here, the information for NF_3 market distribution in 2018 is taken from m.huaon.com (39); the distribution of LCD production capacity in 2018 from statista.com (22); the information on wafter capacity for semiconductor in 2015 is taken from Platzer et al. (23); the regional emissions are redistributed to each grid cell based on the intensity of night light data in 2016 from NASA (https://eoimages.gsfc.nasa.gov/images/imagerecords/144000/1

44897/BlackMarble_2016_01deg_gray_geo.tif). Night lights data represent both population and industrialization density and were used for approximating the prior distribution of trace gases released from industrial activities in a former study. (40) The final proportions for each region used in this study are listed in the middle column in Supplementary Table 2. We assume constant emissions in each grid throughout the year.

The 11 regions in the state vector are North America; Europe; Africa; central and south America; Oceania; South Asia (India, Sri Lanka, Pakistan and Bangladesh); south China (with Taiwan included); South Korea; Japan; southeast Asia; northeast China (Supplementary Figure 1). East Asia is separated into South China, Northeast China, South Korea, and Japan, as the sensitivity of mole fraction to the emissions changes in the regions from east Asia at the AGAGE measurement sites is different (Supplementary Figure 15). The rest of the world is not optimized and hence not included in the state vector, as the NF_3 related industrial activity is very low, and the NF_3 emissions are negligible according to current available information (e.g., the maximum NF_3 emission is 0.48×10^{-3} Gg yr^{-1} in 2016 in Russia according to UNFCCC data set, (41) which is negligible comparing to the minimum in other region, such as Japan). We estimate emissions for south China (with Taiwan included), northeast China, Japan, and South Korea separately to account for the NF_3 emissions produced by flat screen and semiconductor factories in these regions. In South Asia, only India, Sri Lanka, Pakistan, and Bangladesh are considered in this region, and the remaining countries are assigned to the rest of the world due to the expectation that there are negligible NF_3 emissions from these regions. Turkey is included in the European domain. China is separated into South China (including Hong Kong and Macao plus Taiwan) and Northeast China.

INVERSION THEORETICAL FRAMEWORK

We optimize the global NF$_3$ emissions based on Bayes' theorem. The Bayesian cost function J of the inverse problem is written as

$$J(x) = (x - x_A) S_A^{-1}(x - x_A) + (y - Kx) S_O^{-1}(y - Kx)$$

in which x is the state vector. x_A is the prior emissions from different regions. Here, both x and x_A are emissions from 11 regions and 108 months (from 2014 January to 2022 December), leading to 1188 unknowns (Supplementary Figure 1). K is the Jacobian matrix of the monthly sensitivity of modeled NF$_3$ mole fractions at the measurement sites to the perturbation (10%) of prior NF$_3$ emissions from the different regions. Here, we track the perturbed emissions for one year and then approximate the remaining sensitivity as a two-year exponentially decay toward the globally well mixed mole fraction, (42) after which, the perturbed emissions are assumed to be well mixed globally. y is the monthly mean difference between the measurements and the mole fractions derived from the simulation with unperturbed prior emissions and thus, x_A is zero, and the posterior emissions represent the deviations from the prior emissions. S_A is the prior state vector covariance, while S_O is the observational and model error covariance. We use 25% of the global annual prior mean emissions as the annual errors (standard deviation) for each region to construct the prior error covariance matrix. 25% is chosen because it is the maximum difference of the proportion information between the semiconductor and LCD in Supplementary Table 2. We assume no spatial correlation between the emissions from different regions and no temporal correlations between different months, so only diagonal elements in the prior error covariance matrix. The observational error covariance matrix S_O is approximated as the standard deviations of monthly mean measurements at each site. Errors of all measurements are assumed to be independent.

By minimizing J, with respect to x, we get the analytical solution for posterior emissions changes:

$$\hat{x} = x_a + G(y - Kx)$$

where G is the gain matrix:

$$G = (K^T S_o^{-1} K + S_A^{-1})^{-1} K^T S_o^{-1}$$

Bottom-Up Estimates

The total bottom-up emissions E (Gg yr^{-1}) is calculated using the following equation:

$$E_{total} = E_{production} + E_{end-use}$$

$E_{production}$ is the NF$_3$ emitted during NF$_3$ production, and $E_{end-use}$ is the aggregated NF$_3$ released during the end use of NF$_3$ from different industrial activities, mainly semiconductor, LCD, and PV cells. They are calculated using the following equations:

$$E_{production} = P_p \times r_p$$

P_p is the amount of production, and $r_p = E_{production}/P_p$ is the production ratio from all the processes relating to of NF$_3$ production.

$$E_{end-use} = \sum_{i=1}^{N}(P \times r_{supply}) \times (1 - r_{use})$$

For Japanese bottom-up emissions (Japan being the only country for which the required data is publicly available), $N = 2$, as there are mainly semiconductor and flat panel display (LCD) industries in this country. P is the total consumed NF$_3$ amount from all industry activities (tons); r_{supply} is the process supply rate (also known as "heel factor"), 90% is deployed here, meaning 10% of the NF$_3$ in the cylinders will not be used (43); r_{use} is the use rate, containing two catalogs: in situ and remote. In-situ NF$_3$ cleaning process is analogous to the process for other cleaning gases like C_2F_6 and C_3F_8. Remote plasma sources dissociate NF$_3$ into fluorine radicals before they enter the chamber. For semiconductors, r_{use} is 80% for in situ, 98% for remote; for LCDs, r_{use} is 70% for in situ, 97% for remote. (43) Since the use rate is much lower for NF_3 in situ use, emissions are typically much higher as more unused NF_3 exists in the tools used in manufacturing processes. This can be mitigated by abatement, that is destruction of NF_3 in the waste gas stream. We do not consider abatement for $E_{end-use}$ here.

For global bottom-up emissions, N = 3, as production of certain PV cells can also results in NF_3 emissions. We assume that the parameters (r_{use}, r_p) used for Japan also applies to the global scale. The global NF_3 production data and global demand data between 2017 and 2020 are from m.huaon.com, (28) and the global demand data in 2015 and 2016 are from Adams. (29) The global production in 2015 and 2016 are inferred from our top-down emissions divided by the scaled emission ratio from our study and the value in 2011 from Arnold et al. (6) Here, the proportion of NF_3 consumption used for PV cells among global demand is using 3%, with a total leaking ratio 0.017. (8) The global NF_3 consumption in semiconductor was taken from World Semiconductor Council, (44) while the amount used in LCD is the remaining in the global demand. The calculated total bottom-up emissions for in situ and remote are listed in Supplementary Table 8 and the details of the data are in the supporting Excel files.

DATA AVAILABILITY

The observation data from AGAGE are available on https://agage.mit.edu/

Supporting Information

The Supporting Information is available free of charge at https://pubs.acs.org/doi/10.1021/acs.est.4c04507.

- Table S1: Information about measurement stations from AGAGE; Table S2: Regional percentages of factors used for prior NF_3 emissions; Table S3: Performance of the inversion simulations using RMSE; Table S4: Global and regional annual posterior emissions averaged from three inversion runs; Table S5: Global and regional average emissions over the simulation period; Table S6: Global production, NF_3 emission ratio, and their uncertainties; Table S7: Global demand and Chinese (mainland) demand data; Table S8: Bottom-up estimated emissions for Japan and other regions; Figure S1: Regions used in the inversion; Figure S2: Annual global and regional posterior NF_3 emissions from three inversion runs; Figure S3: Annual global and regional prior and posterior NF_3 emissions from the average of the three inversion runs; Figure S4: Error reduction from the average of the three inversion simulations; Figure S5: Correlations between optimized emissions from the eleven regions; Figure S6: Observed and modeled monthly mean NF_3 mole fractions using the prior and posterior estimates at eight AGAGE measurement sites; Figure S7: Differences between observations and GEOS-Chem modeled monthly mean NF_3 mole fractions using the prior and posterior estimates; Figure S8: Sensitivity to different network choices; Figure S9: Sensitivity to different prior emission error; Figure S10: Sensitivity to observational error; Figure S11: Sensitivity to prior emissions magnitude; Figure S12: Sensitivity to prior emissions trend; Figure S13: Sensitivity to constant prior emissions (PDF)

https://pubs.acs.org/doi/10.1021/acs.est.4c04507

Appendix D

From the European Commission, Climate-friendly alternatives to HFCs

To avoid the use and emissions of hydrofluorocarbons (HFCs), a variety of climate-friendly, energy-efficient, safe and proven alternatives are available.

Due to different thermodynamic and safety properties of the alternatives, there is no 'one size fits all' solution. The suitability of a certain alternative must be considered separately for each category of product and equipment; and in some cases also take into account the geographical location in which the product and equipment are used.

Why use alternatives?

The climate impact of a substance is commonly expressed as the global warming potential (GWP). The lower the GWP, the more climate-friendly the substance.

HFCs have a very high GWP and are hence potent greenhouse gases. Most of the HFCs are used as refrigerants in refrigeration and air conditioning

(RAC) equipment, but also as blowing agents, aerosol propellants and solvents.

To mitigate emissions of substances with a high GWP and comply with the [F-Gas Regulation](), each sector needs to find solutions to quickly switch to low GWP refrigerants.

Alternatives and safety groups

Most of the HFCs are used as refrigerants in refrigeration and air conditioning (RAC) equipment, but also as blowing agents, aerosol propellants and solvents.

In the following, alternatives to commonly used HFCs are listed for different sectors.

The alternatives include

- Natural refrigerants
- HFCs with lower GWP, such as R32
- Hydrofluoroolefins (HFOs)
- HFC-HFO blends.

Commercial refrigeration

Commercial refrigeration applications include stand-alone equipment, condensing units and centralised systems.

Plug-in equipment used in small stores and supermarkets, such as vending machines relying on hydrocarbons, has become available in recent years throughout the world. CO_2-based systems have also been introduced.

In large refrigeration systems for supermarkets ('centralised systems'), CO_2 cascade systems are an alternative to commonly used HFC systems in many climates.

The cascade refrigeration system is structured by connecting at least two refrigeration systems in series, with a higher-temperature side and a lower-temperature side. In these cycles different refrigerants are used, which have different freezing and boiling points. This is more efficient compared to conventional refrigeration systems.

Hydrocarbons have also proven to be highly efficient alternatives in most applications under high ambient temperatures, except for larger condensing units.

Industrial refrigeration

In industrial refrigeration, such as large cooling facilities for food processing or process cooling in the chemical industry, ammonia systems have been used for many years.

Ammonia has been the most popular replacement option to R404A and its use is already widespread.

In Europe, but also in other parts of the world such as North America, an increasing number of cascade systems with ammonia and CO_2 have been installed in the food and beverage industry.

Stationary air conditioning

Stationary air conditioning (AC) is designed to control the thermal comfort of living and working rooms. The stationary AC sector can be broken down into several sub-categories:

- Moveable room AC:
  Devices that are hermetically sealed and can be moved between rooms by the user. Mostly used in private households.
- Single split AC:
  System that consists of one outdoor and one indoor unit linked by refrigerant piping, needing installation at the site of storage. Predominantly used in private households.
- Multi split AC/VRF:
  System that consists of one outdoor unit and multiple indoor units. Further developed systems enable a variable refrigerant flow (VRF) towards every indoor unit. Used in commercial facilities.
- Chiller:
  System in which the refrigerant cools down a liquid (normally water) that is then circulated to cool air in commercial or industrial facilities.
- Heat pump:
  System that is able to provide heating or cooling by

transferring heat from or to an external reservoir (such as the ground, water or outside air). Used both in private households and commercial facilities.

In room air conditioning systems, hydrocarbons are safely used as alternative refrigerants in several countries such as India and China, but they are not yet common in the EU.

In chillers, hydrocarbons and ammonia are safe and energy-efficient alternatives to HFCs, both under moderate and high ambient temperature conditions. Heat pumps are also used with hydrocarbons, additionally CO_2 is available on the market.

Mobile air conditioning

The refrigerant R134a used in the air conditioning of cars is prohibited in new cars thanks to EU Directive 2006/40/EC on mobile air-conditioning systems (the 'MAC Directive').

The main substitute is the R1234yf, which is almost exclusively used. The only alternative to this is CO_2, which is currently used by some car manufacturers and expected to become more widespread in the future.

CO_2 is also expected to become available as an alternative in the future for duty vehicles, buses and trains.

Transport refrigeration

Lately, R448A, R449A and R452A have become quite common to replace R404A in road transport refrigerated vehicles. R452A has a very high GWP of 2140 and hence will not be suitable for future use. For refrigerated containers, CO_2 can be used as a long-term alternative.

Foam blowing

Polyurethane (PU) foam: Only few PU foam products are still manufactured with HFC blowing agents. The vast majority rely on hydrocarbons such as pentane or cyclo-pentane without loss in energy efficiency. HFCs are mainly limited to on-site application of PU spray foam. For this and some niche applications, unsaturated HFCs are already commercially available.

Extruded polystyrene (XPS): Major manufacturers of XPS insulation boards have already converted their production facilities to organic solvents or HFOs. The remaining users of HFCs are switching to HFOs. Energy efficiency of HFOs is considered to be better than that of HFCs.

Studies

A number of studies on the feasibility and availability of alternatives at sub-sectoral level have been carried out by various renown experts, including an extensive analysis carried out for the European Commission by the independent consultant Öko-Recherche in the context of developing Regulation (EU) No 517/2014.

https://climate.ec.europa.eu/eu-action/fluorinated-greenhouse-gases/climate-friendly-alternatives-hfcs_en

Appendix E

Total U.S. CO2 Greenhouse Gas Emissions in 2022

Total Emissions in 2022 = 6,343 Million Metric Tons of CO_2 equivalent. Percentages may not add up to 100% due to independent rounding. The Industry and Commercial/Residential (including buildings) sectors consume large amounts of electricity. The share of overall emissions in these end-use sectors is significantly higher when including indirect electricity-related emissions. Greenhouse gas emissions from commercial and residential buildings and industry increase substantially when emissions from electricity end-use are included, due to the relatively large share of electricity use (e.g., heating, ventilation, and air conditioning; lighting; and appliances) in these sectors. Land Use, Land-Use Change, and Forestry in the United States is a net sink and offsets 12% of these greenhouse gas emissions. This net sink is not shown in the above diagram. All emission estimates are sourced from the Inventory of U.S. Greenhouse Gas Emissions and Sinks: 1990–2022.

https://www.epa.gov/ghgemissions/sources-greenhouse-gas-emissions#:~:text=Sector%20in%202022-,Total%20Emissions%20in%202022%20%3D%206%2C343%20Million%20Metric%20Tons%20of%20CO%E2%82%82,of%20these%20greenhouse%20gas%20emissions.

Appendix F

UC Irvine 10/29/2008 report on NF3

September 29, 2008

The missing greenhouse gas

Flat-screen TV manufacturing emits chemical with warming effect

Share on X (Twitter)Share on FacebookShare on LinkedInShare on Email

An unregulated chemical used to make flat-screen televisions and computers has 17,000 times the climate-warming effect of carbon dioxide, say UC Irvine Earth system scientists Michael Prather and Juno Hsu. Their assessment of nitrogen trifluoride, or NF_3, in a recent issue of *Geophysical Research Letters* caught the attention of global warming experts and journalists worldwide, with news stories appearing in the *Los Angeles Times*, *Discover*, *New Scientist*, *Chemistry World* and on National Public Radio.

Prather, Fred Kavli Chair and director of the UCI Environment Institute, discusses NF_3 and why it is such a global warming threat:

Q: What is NF_3 and how does it react in the atmosphere?

A: NF3 is a man-made gas that – once released into the atmosphere – circulates from the surface to the stratosphere hundreds of times before it is destroyed by solar ultraviolet radiation. The average lifetime of an NF3 molecule in the atmosphere is 550 years. NF3 is nearly chemically inert in the atmosphere, but it is very effective in absorbing the infrared radiation that the Earth emits. By trapping this infrared radiation, NF3 becomes a potent greenhouse gas. In terms of kilograms emitted, NF3 is about 17,000 times more effective as a greenhouse gas than carbon dioxide.

Q: How is NF3 used to make flat-screen televisions and computers?

A: When hit with microwave discharge or a plasma beam, NF3 releases fluorine atoms that are used to clean the chamber in which flat-screen liquid crystal display panels are made. It also can be used to release fluorine to etch and cut the silicon substrate for computer chips. As I understand it, NF3 is used in large volumes in the LCD screen process.

Q: Why do you consider NF3 the "missing greenhouse gas"?

A: NF3 is not included in the Kyoto Protocol list of greenhouse gases. This fact is perhaps an oddity. NF3, like many synthetic greenhouse gases, was not recognized specifically as a major industrial gas in 1995 when the United Nations' Intergovernmental Panel on Climate Change report listed the global warming potentials of many greenhouse gases. The rapid rise in production by the chemical industry had gone almost unnoticed.

Q: Is UCI measuring NF3, and if so, how?

A: The labs of <u>Eric Saltzman and Murat Aydin</u> in Earth system science are preparing to measure the atmospheric abundance of NF3. It is an extremely slippery molecule. There are probably only three laboratories in the world with the experience of measuring gases like NF3, and UCI has one of them.

Q: How should industry address the NF3 issue?

A: The major NF3 producer, Air Products and Chemicals, Inc., says that 98 percent of the gas is destroyed during the manufacturing process, and thus NF3 is an environmentally safe gas. However, there should be no doubt that NF3 is a highly hazardous gas in terms of global warming. We have no reliable estimates about leakage during production, shipping and decommission. This gas is extremely volatile and difficult to measure at low abundances, so we cannot be sure that industry estimates or measurements are accurate.

Q: What can the general public do about NF3?

A: The public should recognize that technology has a price and that the manufacture of these screens is not greenhouse-free. NF3 is not special; it is one of many high-tech products that potentially can have a large carbon footprint. The public should demand that the carbon footprint of such products be evaluated independently (not by the chemical manufacturers) and disclosed as energy efficiency is posted on appliances.

Appendix G

From Science Direct, Carbon Tetrafluoride, based on Chemical Engineering Journal, 2021

Exposure and Exposure Monitoring
Airborne fluoride exists in gaseous (hydrogen fluoride, carbon tetrafluoride (CF_4), hexafluoroethane (C_2F_6), and silicon tetrafluoride) and particulate forms (cryolite, chiolite ($Na_5Al_3F_{14}$), calcium fluoride, aluminum fluoride, and sodium fluoride) emitted from both natural and anthropogenic sources. Fluorides are normally found in very small amounts in the urban air <1 µg m^{-3}.
Below pH 5, fluorine is almost entirely complexed with aluminum and consequently, the concentration of free fluorine is low. As the pH increases, Al–OH complexes dominate over Al–F complexes and the free fluorine level increases. Fluoride levels in surface waters vary according to location and proximity to emission sources. Surface water concentrations generally range from 0.01 to 0.3 mg l^{-1}, while seawater contains more fluoride than freshwater, with concentrations ranging from 1.2 to 1.5 mg l^{-1}.
Fluoride levels in terrestrial biota tend to be increased in areas with high fluoride levels due to both natural and

anthropogenic sources. Fluoride is found in most types of soil, ranging from 20 to 1000 µg g^{-1} in areas without natural phosphate or fluoride deposits and up to many thousand micrograms per gram in mineral soils with deposits of fluoride. Airborne gaseous and particulate fluorides tend to be deposited within the surface layer of soils and the soluble fluoride content is biologically important to plants and animals.

While food generally contains low levels of fluoride, it can be higher in areas where phosphate fertilizers have been used. The average daily fluoride intake by adults from food and water ranges from 1 to 2.7 mg depending on whether the water is fluoridated.

2.2 Fluoride in air

Airborne fluoride exists in gaseous and particulate forms emitted from both natural and anthropogenic sources. The gaseous fluorides include hydrogen fluoride (HF), carbon tetrafluoride (CF_4), hexafluoroethane (C_2F_6) and silicon tetrafluoride (SiF_4), while particulate fluorides include cryolite (Na_3AlF_6), chiolite ($Na_5Al_3F_{14}$), calcium fluoride (CaF_2), aluminium fluoride (AlF_3) and sodium fluoride (NaF) [17]. The distribution and deposition of airborne fluoride are dependent upon emission strength, meteorological conditions, topography, particle size and chemical reactivity [22,23].

The annual global release of hydrogen fluoride from natural sources, through degassing and explosive eruptions, ranges from 0.06 to 6 Tg, of which 0.05–5 Tg is released into the troposphere and 0.005–0.5 Tg is erupted explosively into the stratosphere [15]. Thus, volcanoes must be regarded, next to the photolysis of anthropogenic halocarbons in the stratosphere, as a significant source of tropospheric and stratospheric HF. Another important natural source of fluoride is sea salt released from the oceans in the form of marine aerosols, containing ~0.02 Tg annually [24]. Anthropogenic sources of fluoride include fossil fuel combustion and industrial waste. Hydrogen fluoride is water soluble and emissions are readily controlled by acid gas scrubbers. HF emission from coal combustion, that is considered to be the main anthropogenic source of HF, was estimated to be 0.18 Tg annually; emission of HF from the combustion of petroleum and natural gas is almost certainly negligible [24]. Apparently only limited data are available concerning total annual emissions of HF from industrial operations; however, there is evidence that emissions of fluorides have been declining [24,25].

Fluorine-containing gases do not disperse so rapidly in the atmosphere as other gases. Smoke is often carried over great distances and can cause considerable damage when it finally descends. Dässler and Grumbach [26], for instance, were able to detect fluorine-containing gases emitted under very humid conditions at a distance of 1–2 km from the source, and reported that fluorine levels at a distance of 7–8 km from the source could be higher than in nearby zones. Molecular fluorine and hydrogen fluoride are largely responsible for the plant damage caused by stack gases [18]. The effects of fluoro-organics on plants and organisms are presented in details in Vol. 1 of this series by Davison and Weinstein [27]. Hydrogen fluoride is reportedly 1000 times as harmful as sulphur dioxide and there can be no doubt that the harmful effect of the latter on plants is greatly exacerbated by traces of fluorine [28].

Fluoride concentrations in ambient air were studied by Thompson et al. [29] from 1966 to 1968. A total of 2164 samples were taken in non-urban and 9175 in urban areas. 98.5% of those from non-urban areas contained less than 0.05 $\mu g/m^3$, 1.3% contained 0.05–0.09 $\mu g/m^3$ and only 0.1% contained 0.1–0.99 $\mu g/m^3$ of fluoride. Levels in urban areas were slightly higher, 87.8% of samples containing less than 0.05 $\mu g/m^3$, 4.2% containing 0.05–0.09 $\mu g/m^3$, 7.7% containing 0.10–0.99 $\mu g/m^3$ and 0.2% more than 1.00 $\mu g/m^3$ of fluoride. The maximum concentrations observed were 0.16 from non-urban and 1.89 µg/m³ from urban locations.

Climate and Chemicals

As the impacts of climate change grew more apparent, many NGOs working on the toxic impacts of coal and oil came together to form the Climate Action Network (CAN) (http://www.climatenetwork.org). It became clear that some pollutants not only were toxic but also were greenhouse gases, e.g., perfluorobutane (PFB) and carbon tetrafluoride used in the semiconductor industry and produced as a byproduct of aluminum production. Both have been found in the off-gases of overheated Teflon cookware. The CAN is now a worldwide network of over 700 NGOs in more than 90 countries, working to promote government and individual action to limit human-induced climate change to ecologically sustainable levels.

In 2011, UNEP acknowledged that chemical reform needed to be undertaken in the context of the growing interaction of climate change with chemical releases, transport, degradation, exposure, and toxicity. (*Climate change and POPs: Predicting the Impacts*, Report of the United Nations Environment Program (UNEP)/Arctic Monitoring and Assessment Programme (AMAP) Expert Group, January 2011. Available at: http://chm.pops.int.) The report by UNEP and the Arctic Monitoring and Assessment Programme (AMAP) Expert Group, Climate Change and POPs: Predicting the Impacts, concluded that higher temperatures increase primary emissions and secondary releases of POPs from materials, products, and stockpiles. The evidence of increased remobilization of POPs and heavy metals from glacial and permafrost melt is a significant blow to indigenous peoples already under stress from transboundary POPs. Temperature was also shown to affect POP toxicity, and other climate change impacts on salinity, ocean acidification, eutrophication, and water oxygen levels could (either alone or in combination) enhance the toxic effects of POPs. IPEN's cochair was a coauthor of the report and the network has helped disseminate its findings.

12.5.2 Aluminum Smelter Emission Control

Emission problems of conventional aluminum smelters center on fluoride losses, which preabatement (before ca. 1972) was at the rate of some 21 kg/tonne of aluminum produced (Table 12.7). The bulk of this fluoride loss occurred from the operating electrolytic cells, and two-thirds or more of this was gaseous fluoride. The chief constituents of the fluoride discharge were cryolite (Na_3AlF_6), aluminum fluoride, calcium fluoride, chiolite ($Na_5Al_3F_{14}$), silicon tetrafluoride, and hydrogen fluoride. A rough guide as to mass discharge rates may be obtained from their consumption rates per unit of aluminum produced (Table 12.7). Most of these substances are lost as fumes or vapors, except for hydrogen fluoride gas, which results from the reaction of traces of moisture present in alumina added to the cell (Eq. 12.30).

Impact of Climate Change

Since 1990, every 5 years a group of approximately 2000 natural and physical scientists, economists, social scientists, and technologists assemble under the auspices of the United Nations-sponsored Intergovernmental Panel on Climate Change (IPCC). These scientists spend 3 years reviewing all of the information on climate change and produce a voluminous report following a public review by others in the scientific community and by governments. "Climate Change 2001" is 2665 pages long and contains more than 10,000 references. Information for this article is based on this report, the references in it, as well as recent studies to ensure accurate, up-to-date information.

Climate is defined as the 30- to 40-year average of weather measurements, such as average seasonal and annual temperature; day and night temperatures; daily highs and lows; precipitation averages and variability; drought frequency and intensity; wind velocity and direction; humidity; solar intensity on the ground and its variability due to cloudiness or pollution; and storm type, frequency, and intensity.

Understanding the complex planetary processes and their interaction requires the effort of a wide range of scientists from many disciplines. Solar astronomers carefully monitor the intensity of the sun's radiation and its fluxuations. The current average rate at which solar energy strikes the earth is 342 watts per square meter (W/m^2). It is found that 168 W/m^2 reaches the earth's surface mostly as visible light: 67 W/m^2 is absorbed directly by the atmosphere, 77 W/m^2 is reflected by clouds, and 30 W/m^2 is reflected from the earth's surface.

A number of trace gases in the atmosphere are transparent to the sun's visible light but trap the radiant heat that the earth's surface attempts to emit back into space. The net effect is that instead of being a frozen ball averaging $-19°C$, Earth is a relatively comfortable $14°C$. This difference of $33°C$ arises from the natural greenhouse effect. Human additions of greenhouse gases appear to have increased the temperature an additional $0.6\pm0.2°C$ during the 20th century. The transmission of visible light from the sun and the trapping of radiant heat from the earth by gases in the atmosphere occur in much the same way as the windows of a greenhouse or an automobile raise the temperature by letting visible light in but trap outgoing radiant heat. The analogy is somewhat imperfect since glass also keeps the warm inside air from mixing with the cooler outside air. The temperature rise at the earth's surface is directly proportional to a measure of heat trapping called radiative forcing. The units are watts per square meter, as for the sun's intensity. The radiative forcing from human additions of carbon dioxide since the beginning of the industrial revolution is 1.46 W/m^2. Methane and nitrous oxide additions have provided relative radiative forcings of 0.48 and 0.15 W/m^2, respectively. Other gases have individual radiative forcings of less than 0.1 W/m^2. The total radiative forcing of all greenhouse gases added by human activity to the atmosphere is estimated to be 2.43 W/m^2. This should be compared to the 342 W/m^2 that reaches the earth from the sun. Hence, the greenhouse gases added by human activity are equivalent to turning up the sun's power by an extra 0.7%, which is enough to cause the global temperature to increase by the observed 0.6°C (1.1°F).

Carbon dioxide is removed from the atmosphere by dissolving in the ocean, by green plants through photosynthesis, and through biogeological processes, such as coral reef and marine shell formation. Approximately half of the carbon dioxide released today will remain in the atmosphere for a century, and one-fourth may be present 200 years from now. Some of the synthetic, industrial greenhouse gases have half-lives of hundreds, thousands, and, in at least one case (carbon tetrafluoride), tens of thousands of years. These long residence times are effectively irreversible changes in the composition of the atmosphere and in climate parameters on the timescale of human generations. This is why some argue that the continued release of greenhouse gases is not sustainable since it can adversely affect future generations. Even without complete information about the consequences of increasing greenhouse gases for climate change, many invoke the precautionary principle and urge the world to move toward reducing and eliminating them. The post-ice age period in which human civilization has evolved is known as the Holocene. It has long been assumed that this represented an extended, stable, and generally warm climate period. Temperatures were slightly higher than those of today at the beginning of the Holocene and gradually declined until recently. Measurements from Greenland have shown that the general changes were subject to sudden shifts in temperature on timescales of a decade or a century. It is believed that these rapid temperature changes occurred as the result of sudden alterations in oceanic and <u>atmospheric circulation</u>. Direct land and ocean instrumental temperature measurements began in 1861. These records show two periods of global temperature rise, from 1910 to 1945 and from 1976 to the present. In between, from 1946 to 1975, there was little change in global average temperature, although some regions cooled while others warmed. The total rise for land-based measurements from 1861 to 2000 is estimated by the IPCC to be $0.6°C \pm 0.2°C$. The warming rate across the entire earth's surface during the two periods has been $0.15°C$ per decade. Sea-surface temperatures are increasing at approximately one-third this rate, and it has been found that the heat content of the top 300 m of the ocean increased significantly during the second half of the 20th century. The lower atmosphere up to 8 km has also warmed, but at approximately one-third the land rate. Daily minimum temperatures are rising at $0.2°C$ per decade, approximately twice the rate of daily maximum temperatures. This has extended the frost-free season in northern areas, and indeed satellite data confirm that total plant growth is increasing at latitudes above 50°N.

There has been much discussion of whether the rise in the instrumental record during the 20th century is simply a return from a <u>little ice age</u> that occurred between the 15th and 19th centuries to conditions of a medieval warming period that predated it. An analysis of surrogate measurements of temperature utilizing data from <u>tree rings</u>, coral reefs, and ice cores that extends back 1000 years clearly shows that the temperature rise of the 20th century for the Northern Hemisphere lies outside the variability of the period. In fact, the current rate of increase, $0.6°C$ per century, reverses a downward trend of

−0.02°C per century during the previous 900 years. The statistical analysis clearly shows that there is less than a 5% probability that the recent temperature increase is due to random variability in the climate system.

The IPCC concludes that the 20th century was the warmest in the past 1000 years, and the 1990s were the warmest decade in that period. The year 1998 may have been the warmest single year, 0.7°C above the 40-year average from 1961 to 1990 and nearly 1.1°C warmer than the projected average temperature extrapolated from the previous 900-year trend.

By the end of the 19th century, scientists had identified the major greenhouse gases in the atmosphere. In 1897, Svente Arhenius, a distinguished Swedish chemist, carried out the first calculation of the effect of greenhouse gases on the temperature of the earth. His estimate, made without a computer or even a good calculator, was surprisingly close to what we know today. Arhenius also performed a thought experiment in which he recognized that the industrial revolution was moving carbon from under the ground in the form of coal and oil and putting it into the atmosphere as carbon dioxide. What would happen, he asked, if human activity were to double the concentration of carbon dioxide in the atmosphere? His answer was approximately twice the upper limit of the range estimated by today's best climate models of 1.5–4.5°C. This is called the climate sensitivity. The actual increase in temperature depends on the amount of greenhouse gas released into the atmosphere. It is estimated that this will amount to approximately a doubling of carbon dioxide equivalent from preindustrial levels some time before 2100. Systematic measurements of atmospheric composition begun in 1959 by Charles David Keeling in Hawaii have confirmed that carbon dioxide is indeed increasing along with the increase in fossil fuel consumption. Air trapped in glacial ice confirms that carbon dioxide in the atmosphere has risen one-third from

280 parts per million (ppm) in preindustrial times to 373 ppm in 2002.

Human activities are influencing climate in several ways. The most direct is through the release of greenhouse gases. The greatest share of greenhouse gases comes from the combustion of fossil fuels, such as coal, oil, and natural gas, which release carbon dioxide when burned. Coal releases nearly twice as much carbon dioxide as does natural gas for each unit of heat energy released. Oil releases an amount in between.

Methane released from coal mining constitutes the largest source of anthropogenic methane. Methane releases during natural gas leaks during drilling and transportation are also significant. Long-burning coal mine and peat fires also contribute significant amounts of carbon dioxide, methane, and many air pollutants to the atmosphere.

Land clearance, deforestation, and forest fires release carbon dioxide (23% of total human emissions) and methane.

Land-use change alters the reflectivity or albedo of the land. Deforestation in the northern forests usually increases reflectivity by replacing dark evergreen trees with more white-reflecting snow. In the tropics, less solar energy is absorbed in most cases after forests are removed. Urbanization usually increases the absorption of solar energy. The net effect of land-use change is thought to offset warming by approximately -0.2 ± 0.2 W/m^2.

Combustion of fossil fuels and clearing of land also contribute dust, particulates, soot, and aerosol droplets. These either reflect sunlight away from the earth or absorb radiation. Some aerosols are also believed to increase the amount of cloud cover.

Agriculture releases carbon dioxide through fossil fuel energy use and from the oxidation of soils. This sector is the major producer of nitrous oxide from the bacterial breakdown of nitrogen fertilizer. Rice culture and livestock are also major producers of methane.

Industrial processes release a variety of greenhouse gases into the atmosphere either during manufacturing or through the use of products.
Air pollution that results in ozone and other gases near the earth's surface traps heat, whereas the depletion of the stratospheric ozone in the upper atmosphere allows more heat to escape to space. The pollution near the earth's surface is the larger effect.
The waste sector releases large quantities of methane from sewage and industrial wastewater, animal feedlots, and landfills.

The advantage of using NF_3 as an etchant over traditional carbon based etchants such as carbon tetrafluoride (CF_4) and hexafluoroethane (C_2F_6) includes high etching rate, high selectivity for nitride-over-oxide etching and single crystal silicon over thermally grown silicon oxide, and the production of only volatile reaction products, resulting in an etching without carbon based polymer or fluoride residues. Enhanced plasma or thermal cleaning of chemical vapor deposition (CVD) chambers is also a use of NF_3. Residual coating such as silicon are deposited on the internal surfaces of CVD chambers during deposition processes. A plasma using NF_3 can remove these deposits as volatile fluoride like SiF_4 at the process temperature, eliminating the need to remove the internal CVD chamber components to be cleaned by acid tank immersion. At present, large amounts of NF_3 is consumed as a dry etchant and a cleaning gas for the CVD chamber by the electronic industry in Japan.

Hydrophobic membranes have very few or no functional groups which can take part in surface modifications. A variety of materials have been tested for MD membranes, including both polymeric and inorganic materials. The former, being more versatile, are predominantly used. One of the novel membrane surface modification techniques is called plasma treatment and has the potential to significantly alter surface roughness and functional groups to change the membrane's wettability. Oxygen, argon, and carbon tetrafluoride (CF_4) have been applied on membranes to achieve superhydrophobicity (Barati Darband et al., 2018). Plasma micro-nanotexturing can create random and ordered rough surfaces (Ellinas et al., 2017). An example of changes in surface morphology involving PTFE membranes after applications of plasma treatments was presented for MD (Dumée et al., 2016). Table 4 summarizes studies regarding coating of hydrophobic membranes to achieve superhydrophobicity.

Researchers have successfully used the NTP technology to treat wastewater mainly on the laboratory scale. The wastewater volume is usually less than 1000 mg L^{-1}, and the input energy is limited to about 20 kV to cut operational costs (X. Li et al., 2020). However, to realize the full potential of the NTP technology, there is a need to design and implement pilot and industrial-scale plasma reactors with excellent efficiencies to tackle natural wastewater. Dobslaw and Glocker (2020) developed a pilot-scale DBD system. The reactor was combined with a bio-scrubber and absorber to remove volatile organic compounds. As a result, the removal of carbon tetrafluoride-polluted air reached a 100% removal rate. Fröhling et al. (2018) utilized a pilot-scale non-thermal microwave-driven plasma discharge (Plexc²®) in the washing process of endive lettuce to control microbial growth. The NTP-treated water washed the lettuce and achieved a total viable count reduction of 1.8 log units. Andrasch et al. (2017) touted the advantages of Plexc²® microwave plasma processed water (PPW) and its potential use for sanitary purposes in the food industry. The researchers scaled up a pilot plant to generate PPW to deactivate the microbe *Cichorium endivia* from freshly cut lettuce.

Poly- and perfluoroalkyl substances (PFAS) (~10–10^5 ng/L) were collected from 13 different sites and treated using a mobile pilot-scale CD plasma reactor by Singh et al. (2019a). The reactor removed numerous PFAS from the contaminated water with varying efficiencies ranging from 36 to 99%. The NTP technology has proven efficient in degrading trace organic compounds such as endocrine-disrupting compounds and pharmaceuticals. Gerrity et al. (2010) implemented a pilot-scale CD plasma reactor for that purpose and achieved promising results. The authors concluded that the NTP technology was a viable alternative for contaminant degradation. Another group of researchers designed a pilot-scale pulsed corona NTP reactor for hospital wastewater treatment. The reactor achieved an 87% reduction in the 32 detected pharmaceuticals in the wastewater (Ajo et al., 2018). The NTP technology has to undergo several efficiency and performance tests before it can be fully implemented for commercial use, considering the operational cost and durability, because undoubtedly, the technology has been demonstrated to be highly effective in abating contaminants in wastewater.

The future of the commercialization of the NTP technology indirectly relies on the implementation of pilot projects and industrial applications. Published articles on pilot-scale NTP projects are limited, inadvertently affecting the realization of the technology on a commercial scale. Ideally, the NTP technology is combined with other conventional treatment methods as a tertiary alternative to abate recalcitrant pollutants that escape the conventional treatment process. The NTP technology is highly versatile and can be adjusted to target specific pollutants. For example, the generation of reactive species can be regulated by changing the working gas, and exogenous catalysts can be added to improve degradation efficiencies. The degradation intermediates can serve as a valuable source for a resource recovery to replace raw materials. The potential of the NTP technology is vast, and implementing pilot-scale projects is essential to understand the treatment dynamics of natural wastewater.

2.2 Thermal treatments

Thermal treatment of PFAS includes combustion treatment where oxygen and high temperature are required, and non-combustion thermal treatment where low or no oxygen and relatively low temperature are used.

2.2.1 Combustion treatment

PFAS contaminated solids, including industrial and municipal solid wastes, spent activated carbon and anionic exchange resins, contaminated soils, and <u>sewage sludge</u>, are generally treated thermally within <u>oxidizers</u>, <u>combustors</u>, and incinerators designed for the purpose of organic waste destruction (Krug et al., 2022). Incineration has been used for destruction of halogenated organic chemicals, such as refrigerants and ozone depleting substances at high temperatures and long residence times by breaking carbon-halogen bonds, after which alkali scrubbing is used to remove halogen from <u>flue gas</u> (Oppelt, 1987). PFAS (including fluorinated refrigerants) are halogenated organic chemicals that are most difficult to incinerate, because the C—F bond is at least 50 % stronger than those of other carbon-halogen bonds (O'Hagan, 2008). Furthermore, the flame-sustaining free radical chain reactions can also be terminated by fluorine due to its electronegativity and reactivity. Hence, incomplete combustion of fluorinated product could be emitted from thermal oxidizers, combustors and incinerators.

In general, halogenated organic compounds are thermally decomposed via unimolecular and bimolecular reactions with flame radicals. Fluorinated organic compounds require much higher temperatures to achieve 99.99 % destruction than do their other halogenated counterparts under the same conditions. Carbon tetrafluoride (CF_4) might be generated in the incineration process and is the most difficult fluorinated organic compound to decompose based on calculated bond energies, requiring temperatures over 1400 °C with 1 s gas phase residence time to achieve 99.99 % destruction efficiency (Krug et al., 2022).

2.2.2 Non-combustion treatment

Pyrolysis/gasification as shown in Fig. 2, is one of the most promising non-combustion treatments for commercial destruction of PFAS, especially PFAS in biosolids, which was identified by the PFAS Innovative Treatment Team in the USA (Berg et al., 2022). Pyrolysis is a process that decomposes materials at moderately elevated temperatures in an oxygen-free or low oxygen environment. Gasification is similar to pyrolysis but uses small quantities of oxygen, taking advantage of the partial combustion process to provide the heat to operate the process. The oxygen-free environment in pyrolysis and the low oxygen environment of gasification distinguish these techniques from incineration.

Since PFAS have been widely detected in biosolids from wastewater treatment plants (WWTPs), there are concerns for land-application of biosolids. Pyrolysis and gasification might be used to treat PFAS-contaminated biosolids instead of combustion that is generally used for destruction of PFAS in sewage sludge incinerators.

Since organic wastewater contaminants (hydrocarbons) are dominant in biosolids (Hakeem et al., 2022; Rigby et al., 2021), pyrolysis and certain forms of gasification without oxygen are able to transform biosolids into a biochar and a hydrogen-rich synthetic gas (syngas) via thermal decomposition of C—H bonds of hydrocarbons, and the resultant solids for soil amendment and as a supplemental fuel for biosolids drying operations. It would also reduce energy cost and the emission of greenhouse gas (CO_2) compared to the combustion method, since it operates at low temperature and converts most of the carbon into biochar. Both biochar and syngas can be valuable products. Biochar has many potential applications and is currently used to increase the soil's capacity to hold water and nutrients, requiring less irrigation and fertilizer on crops. Syngas can be used on-site, significantly lowering energy needs. As an additional advantage, pyrolysis and gasification require much lower air flows than incineration, which reduces the size and capital expense of air pollution control equipment.

Pyrolysis/gasification also show promise to reduce PFAS loadings from biosolids without compromising the benefits of the final products, and become an attractive alternative to sewage sludge incineration for reduction of WWTP solids to inert ash, with potential uses as input material in cement production and fine aggregate applications(Bernardin, 2022; Chang et al., 2020).

Pyrolysis or gasification can theoretically vaporise all PFAS and partially destroy some PFAS at its operating temperatures with extended residence times. Residence times vary from minutes to a couple of days as shown in Table 8. Since PFAS transfer into the hydrogen-rich syngas stream, subsequent combustion in a thermal oxidizer (or afterburner) could potentially destroy PFAS. However, the evaluation of potential products of incomplete destruction remains a subject for further investigation and research. The combination of pyrolysis and thermal oxidizer may be more effective at PFAS destruction than some lower temperature sewage sludge incineration processes.

High temperature destruction is usually achieved by combustion processes in the presence of oxygen, and no literature concerning high temperature destruction of PFAS in the absence of oxygen was found. Rather, oxygen has been used to assist combustion of the emitted pyrolysis gas and to raise the temperature above 1000 °C, at which complete mineralisation of PFAS and reaction of fluorine ion with hydrogen occurs. By using syngas combustion, external energy input could be reduced or is not required.

2.2.3 Cost of thermal treatment

Currently, there is no specific cost estimation for PFAS treatment, since it depends on the purpose of treatment. Compared to physical removal/adsorption, PFAS destruction would be significantly more capital and energy intensive.

However, PFAS are considered as 'forever' chemicals, which only can be eliminated by destruction under critical conditions. Thermal treatment is only treatment for PFAS destruction at commercial or pilot scale, and its operation cost can be estimated via existing facilities. For an incinerator with 98–99.99 % efficiency, treating halogenated VOC streams, with a combustion temperature of 1100 °C, residence time of 1.0 s, and use of an acid gas scrubber on the outlet, the cost could be as high as $3600 per metric ton (USEPA, 2006).

2.4 Other chemical and radiation pre- and post-treatments
Recent developments in natural textile dyeing mainly rely on the modification of natural and synthetic fibres using various pre- and post-treatment agents for improvement in color, fastness and functional characteristics of dyed fabrics along with the focus on environmental compatibility of these modification methods. Effect of various surface modification agents containing cationic (Kim et al., 2004; Kim and Park, 2007) and anionic groups (Kamel et al., 2009a, 2011; Bulut and Akar, 2012) on dyeing properties of fabrics with different types of natural dyes have been evaluated frequently. Results indicated that pre-treatment of different types of fabrics with cationic and anionic agents enhances color strength and fastness properties of dyeings over untreated dyeings to a significant extent. Fibre surface modification by the use of synthetic and natural polymers is also employed successfully for enhancing absorption capacity naturally dyed fibres (Janhom et al., 2004, 2006). In recent years, a number of investigations have been carried out to exploit potential applicability of chitosan, a naturally occurring biopolymer with distinct chemical and biological properties, as surface modification agent to improve dyeing performance and biofunctionalisation of cotton (Kavitha et al., 2007; Kim, 2006) and wool fabrics (Dev et al., 2009). The pre-treatment of fabrics with chitosan

increased the binding sites for the dyes resulting in better dye absorption (Chairat et al., 2011).

Recently, dry modification processes such as sputtering and low-temperature plasma (LPT) treatment techniques that generate no waste have been introduced for the pre-treatment and finishing of textile fabrics as environment-friendly alternative to wet chemical applications (Ghoranneviss et al., 2011; Shahidi and Ghoranneviss, 2011). Plasma treatment of fibres improves surface characteristics of fibres while the bulk properties of the fibres are not affected (Barani and Maleki, 2011). Wakida et al. (1998) marked the associated benefits of oxygen, carbon tetrafluoride, and ammonia low-temperature plasma treatments of wool and nylon 6 fabrics for dyeing with several natural dyes, such as cochineal, Chinese cork tree, madder, and gromwell. Park et al. (2008) subjected PET fabrics to chitosan and/or O_2 low-temperature plasma as a pre-mordant in which the fabric was processed by padding, plasma-etching, and immersion procedure and dyed with natural dye of *Caesalpinia sappan*. They showed that pre-treatment of PET fabrics with chitosan and/or plasma is better than a metal mordant in terms of the dye uptake and reduction in dyeing time. Surface pre-activation using air atmospheric plasma and ultraviolet excimer treatment have also been used to activate PET fibrous surfaces yielding hydrophilic species, thereby increasing dyeing affinity with curcumin natural dye (Kerkeni et al., 2012). Chemical modification of cotton fabric with reactive cyclodextrin (R-CD) at different concentrations resulted in remarkable enhancement in printability of cotton fabric as well as color strength of printed sample; however there was no remarkable difference between fastness properties of modified and unmodified samples (Hebeish et al., 2006).

The inherent poor light fastness of several natural dyes is a property which limits their use in modern day applications. The use of UV absorbers and antioxidants are reported to have positive effect on the light fastness of naturally dyed fabrics (Cristea and Vilarem, 2006; Lee et al., 2001). Introduction of a singlet oxygen quenching group into an ultraviolet absorber prepared as a means of improving the light fastness of naturally dyed fabrics provide a novel approach for improving the photostability of natural dyes (Oda, 2012a, 2012b). The photostability of an anionic natural dye can be improved by intercalation into the hydrotalcite layer, if the dye has a hydrophilic nature and a rather planar structure. The intercalated dye is stabilized by the protection from the attack of the atmospheric oxygen (Kohno et al., 2009).

Hydrophobic acrylic fibres pose a big problem for dyers due to their hydrophobic nature. Incorporation of amino groups to acrylic fabrics by treatment with hydroxylamine hydrochloride is reported to increase substantively of the fibre towards some anionic natural dyes (Guesmi et al., 2012a, 2013). Modified acrylic fibre containing different amounts of amidoxime groups showed better dye uptake along with significant improvement in color characteristics (El-Shishtawy et al., 2009). Considering pH sensitivity of natural dyes some researchers implied ammonia after treatment as useful technique to obtain better shades with improved colorimetric properties (Montazer et al., 2004; Montazer and Parvinzadeh, 2004). The preliminary studies on natural dye–surfactant interactions have shown that dyeing of fibres with the natural dyes in the presence of the surfactants enables good shade and fastness in color at low temperatures to be achieved and may minimize the damage to fibres during dyeings (Chandravanshi and Upadhyay, 2012). Use of high energy radiations such as UV (Adeel et al., 2012; Iqbal et al., 2008) and gamma radiation (Bhatti et al., 2010; M'Garrech and Ncib, 2009; Naz et al., 2011) had also been reported to be beneficial in improving the dye uptake and color characteristics of naturally dyed fabrics.

Pre-treatment and finishing of textile fabrics by modern surface modification techniques, such as use of eco-friendly biopolymers, plasma and radiation treatment technologies, relate to environmentally clean textile processing methodology and are much superior to traditional chemical modification methods. Increasing momentum of advanced technologies in sustaining green chemistry aspects of processing and application shows that there is a significant potential to reduce the ecological impact of existing processes with the introduction of these new concepts in real-time methodologies of textile dyeing and finishing focusing on realizing higher productivity gains in consumption of natural dyes without compromising with eco-safety standards.

https://www.sciencedirect.com/topics/earth-and-planetary-sciences/carbon-tetrafluoride

Appendix H

Email query to Sierra Club, EPA, NOAA, University of California, Irvine, climate scientists, University of San Diego, climate scientists and UN Environmental Programme, responses and follow-ups

Email to member.care@sierraclub.org,outreach@noaa.gov,climate-portal@noaa.gov,climate-climatewatch@noaa.gov,amartiny@uci.edu,csorte@uci.edu,Richard Blake,Richard Blake,Another Google,Michael Jurgens,RRB,kmackey@uci.edu,nesdis.pa@noaa.gov,education@noaa.gov,unepinfo@unep.org,executiveoffice@unep.org,JPOCoordinator@unep.org,uneplib@unep.org,unepmedia@unep.org,unep-europe@un.org,unepnyo@un.org,unep-latinamerica-news@un.org,unepap@un.org,unep-westasia@un.org,info@unep-wcmc.org,unep-incplastic-secretariat@un.org,info.switchafricagreen@un.org,communication.roa@unep.org,skimball@uci.edu,thuxman@uci.edu,kprather@ucsd.edu,ckennel@ucsd.edu,mkahru@ucsd.edu,dlubin@ucsd.edu,rsomerville@ucsd.edu,pflatau@ucsd.edu

To Whom It May Concern;

I am currently in the process of writing a book about the "super greenhouse gases." I would very much appreciate a response to these three questions.

Question 1: Are you concerned about the use of "super greenhouse gases," such as nitrogen trifluoride (NF3) which is purported to be 17,200 times a more potent greenhouse gas than carbon dioxide (CO2).

Please explain:

Question 2: Are you concerned that the "super greenhouse gases" might dramatically worsen global warming in the future?

Please explain:

Question 3: If you are in fact concerned, what steps would you recommend be taken to address the issue?

Thanks very much

Sincerely

R. Roy Blake

Initial Response #1, plus attachments:

From Howard Diamond at NOAA

Hello Richard,
Thanks for your inquiry. Well according to the attached paper by Prather and Hsu (2008), nitrogen trifluoride (NF3) can be called the missing greenhouse gas: It is a synthetic chemical produced in industrial quantities; it was not included in the Kyoto basket of greenhouse gases or in national reporting under the United Nations Framework

Convention on Climate Change (UNFCCC); and at the time of that paper, there were no observations documenting its atmospheric concentration. However, since then (and beginning around 2013) it has begun being monitored as noted in the paper by Arnold et al (2013).

In NOAA, we monitor both NF3 at https://gml.noaa.gov/hats/gases/NF3.html as well as SF6 at this link at https://gml.noaa.gov/ccgg/trends_sf6/, as well as host of many other greenhouse gases and not just CO2 and methane. The Global Warming Potential of NF3 is second only to that of SF6 (100-yr GWP = 22,800). So, like any greenhouse gas, we are of course concerned about these, and on an annual basis we publish an Annual Greenhouse Gas Index (AGGI) which documents the increasing amount of heat being added to the atmosphere via the radiative forcing caused by the human-emitted greenhouse gases (GHGs). It is based on the highest quality measurements of GHGs in the atmosphere from sites around the world. I have attached the latest AGGI graphic from 2023 which documents the relative contributions of various GHGs to the radiative forcing load for the planet. At this point, the top 3 GHGs from the standpoint of radiative forcing (and this is a combination of radiative forcing and lifetime in the atmosphere) are CO2, methane (CH4), and nitrous oxide (N2O); with chlorofluorocarbons (CFCs), hydrochlorofluorocarbons (HCFCs), and hydrofluorocarbons (HFCs) rounding out the top 5.

So, in summary, yes, we are concerned with all greenhouse gases and those concerns about NF3 and SF6 are embodied in the monitoring that we do of both of those as well into a host of other GHGs. There is certainly a concern that those gases could worsen the problem, but right now they are not in the top 5 gases in our AGGI index - that does not mean that they are not important, but that their contribution is relatively small at this point in comparison to the big 5 I have noted about. As for your final question, I am not as familiar with the mitigation strategies for NF3 as say for CO2, but I have attached a short whitepaper from Professor F.J. Martin-Torres, from the University of Aberdeen in Scotland, that might be of some assistance to you here, and there are some additional references noted there. Regarding SF6, the US EPA has a link for Best Practices to Reduce SF6 Emissions that you may also find to be of some interest.

Hope that helps, but if you need anything else, please feel free to let me know and I will do my best to assist you.

Regards.

Howard Diamond, PhD
Climate Science Program Manager at NOAA's Air Resources Laboratory
NOTE: This message was **NOT** AI-generated

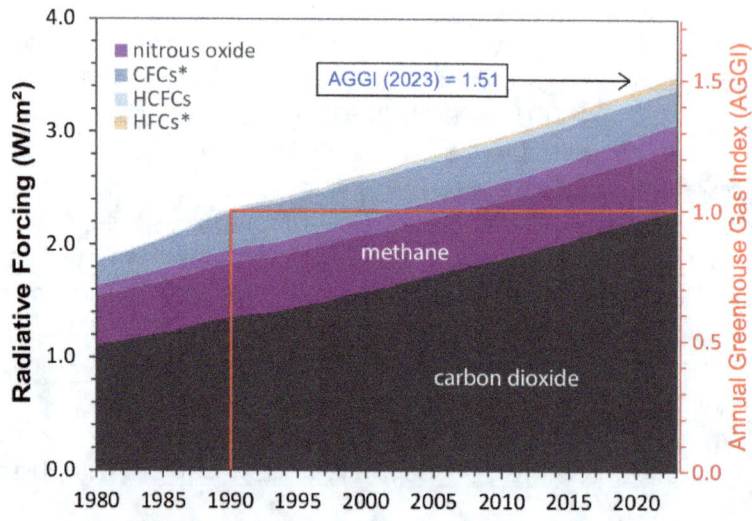

NF3 (Nitrogen trifluoride)
The HATS flask GC-MSD program measures NF3 with the PERSEUS instrument.

Flask NF$_3$ data

Supplemental Info and Data Use Policy

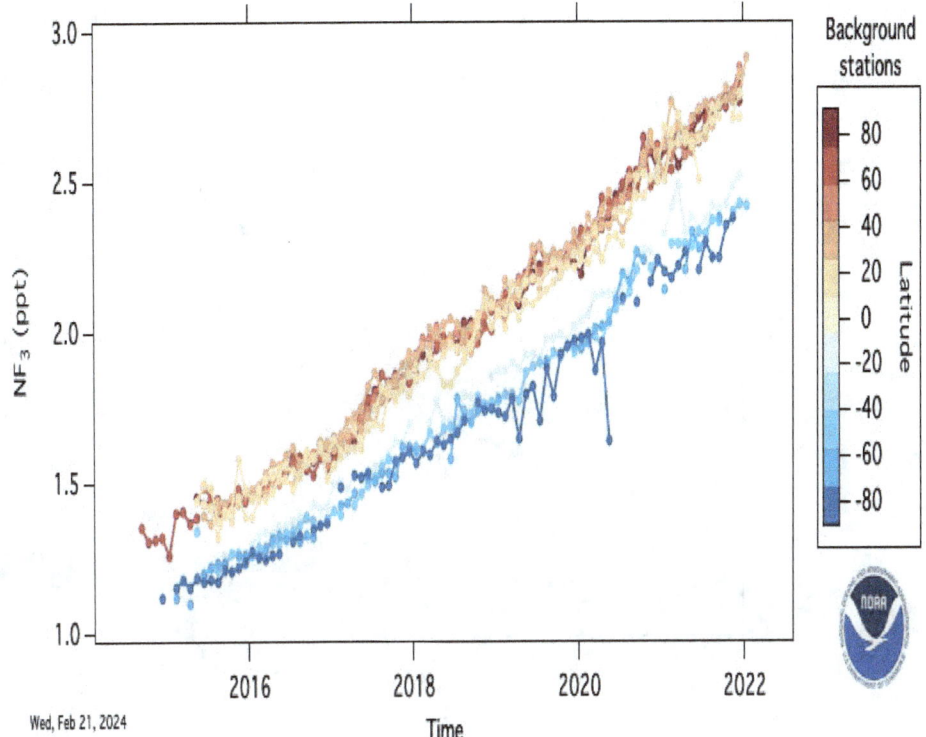

Atmospheric dry mole fractions (parts-per-trillion or ppt) of NF$_3$ measured by GC-MSD in the HATS flask program. Each point represents a monthly mean at one of 8-12 stations: Alert, Canada (ALT), Summit Greenland (SUM), Barrow, Alaska (BRW), Mace Head, Ireland (MHD), Trinidad Head, California (THD), Niwot Ridge, Colorado (NWR), Cape Kumukahi, Hawaii (KUM), Manua Loa, Hawaii (MLO), American Samoa (SMO), Cape Grim, Australia (CGO), Palmer Station, Antarctica (PSA), South Pole, Antarctica (SPO).

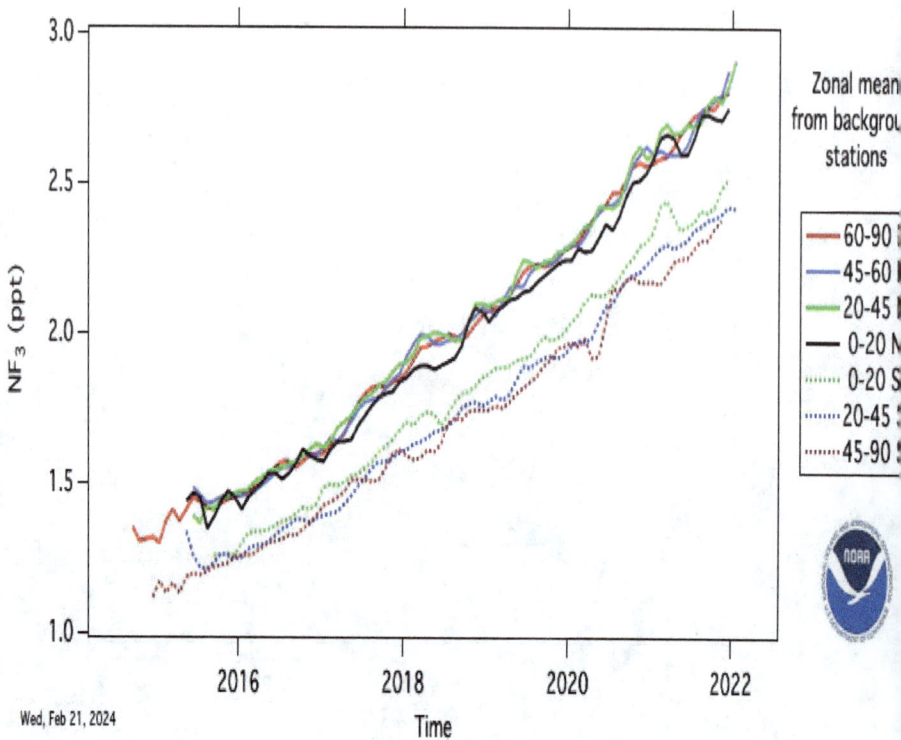

Zonal mean dry mole fractions (ppt) are estimated by equally weighting monthly mean measurements in each latitude band. Small gaps (less than 6 months) are interpolated prior to averaging. Large data outliers are filtered (&gt3 sigma). Solid lines are used for Northern hemispheric measurements and dashed for the Southern hemisphere.

Trends in Atmospheric Sulfur Hexafluoride (SF$_6$)

Global SF$_6$ Monthly Means

June 2024:   11.79 ppt

June 2023:   11.39 ppt

Last updated: Oct 05, 2024

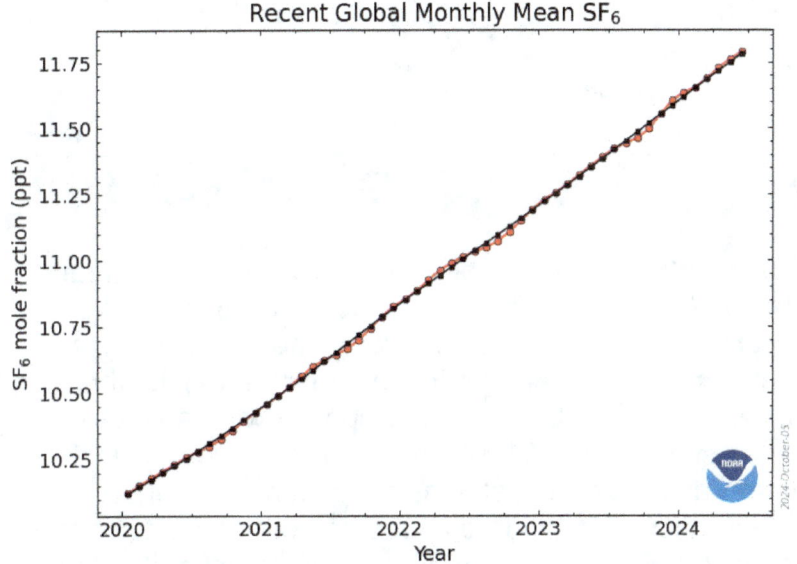

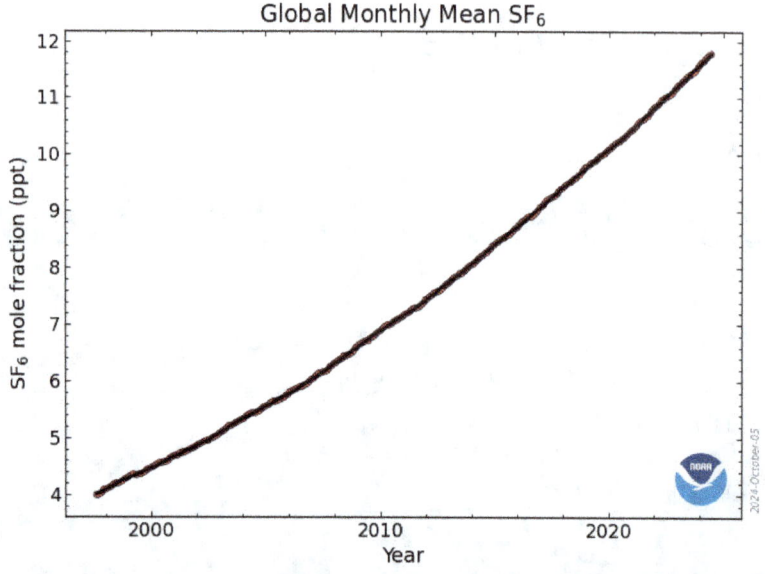

The graphs show globally-averaged, monthly mean atmospheric sulfur hexafluoride abundance determined from marine surface sites. The first graph shows monthly means for the last four years plus the current year, and the second graph shows the NOAA time-series starting in 2001, when we have confidence in the data. Values for the last year are preliminary pending recalibrations of standard gases and other quality control steps. Other impacts on the latest few months of data are described below.

The Global Monitoring Laboratory has measured sulfur hexafluoride since 1997 at a globally distributed network of air sampling sites (*Dlugokencky et al., 1994*). A global average is constructed by first smoothing the data for each site as a function of time, and then smoothed values for each site are fitted as a function of latitude for 48 equal time steps per year. Global means are calculated from the latitude fits at each time step (*Masarie and Tans, 1995*). Go here for more details on how global means are calculated.

Sulfur hexafluoride is reported as a "dry air mole fraction", defined as the number of molecules of sulfur hexafluoride divided by the total number of molecules in the sample, after water vapor has been removed. The mole fraction is expressed as pmol mol^{-1}, abbreviated "ppt" (for parts per trillion; 1 ppt indicates that one out of every trillion molecules in an air sample is SF_6).

In the figures, the red lines and circles are globally averaged monthly mean values centered on the middle of each month. The **black line** and squares show the long-term trend (in principle, similar to a 12-month running mean) where the average seasonal cycle has been removed.

Chlorofluorocarbons (CFCs)

(published in The Chapman & Hall Encyclopedia of Environmental Science, edited by David E. Alexander and Rhodes W. Fairbridge, pp pp.78-80, Kluwer Academic, Boston, MA, 1999.)

James W. Elkins

National Oceanic and Atmospheric Administration (NOAA), Climate Monitoring and Diagnostics Laboratory (CMDL), 325 Broadway, Boulder, CO 80303 U.S.A.

E-mail: James.W.Elkins@noaa.gov, Phone: (303) 497-6224, Fax: (303) 497-6290

Chlorofluorocarbons (CFCs) are nontoxic, nonflammable chemicals containing atoms of carbon, chlorine, and fluorine. They are used in the manufacture of aerosol sprays, blowing agents for foams and packing materials, as solvents, and as refrigerants. CFCs are classified as halocarbons, a class of compounds that contain atoms of carbon and halogen atoms. Individual CFC molecules are labeled with a unique numbering system. For example, the CFC number of 11 indicates the number of atoms of carbon, hydrogen, fluorine, and chlorine (e.g. CCl_3F as CFC-11). The best way to remember the system is the "rule of 90" or add 90 to the CFC number where the first digit is the number of carbon atoms (C), the second digit is the number of hydrogen atoms (H), and the third digit is number of the fluorine atoms (F). The total number of chlorine atoms (Cl) are calculated by the expression: $Cl = 2(C+1) - H - F$. In the example CFC-11 has one carbon, no hydrogen, one fluorine, and therefore 3 chlorine atoms.

Refrigerators in the late 1800s and early 1900s used the toxic gases, ammonia (NH_3), methyl chloride (CH_3Cl), and sulfur dioxide (SO_2), as refrigerants. After a series of fatal accidents in the 1920s when methyl chloride leaked out of refrigerators, a search for a less toxic replacement begun as a collaborative effort of three American corporations- Frigidaire, General Motors, and Du Pont. CFCs were first synthesized in 1928 by Thomas Midgley, Jr. of General Motors, as safer chemicals for refrigerators used in large commercial appilications[1]. Frigidaire was issued the first patent, number 1,886,339, for the formula for CFCs on December 31, 1928. In 1930, General Motors and Du Pont formed the Kinetic Chemical Company to produce Freon (a Du Pont tradename for CFCs) in large quantities. By 1935 Frigidaire and its competitors had sold 8 million new refrigerators in the United States using Freon-12 (CFC-12) made by the Kinetic Chemical Company and those companies that were licensed to manufacture this compound. In 1932 the Carrier Engineering Corporation used Freon-11 (CFC-11) in the worldís first self-contained home air-conditioning unit, called the "Atmospheric

Cabinet".; Because of the CFC safety record for nontoxicity, Freon became the preferred coolant in large air-conditioning systems. Public health codes in many American cities were revised to designate Freon as the only coolant that could be used in public buildings. After World War II, CFCs were used as propellants for bug sprays, paints, hair conditioners, and other health care products. During the late 1950s and early 1960s the CFCs made possible an inexpensive solution to the desire for air conditioning in many automobiles, homes, and office buildings. Later, the growth in CFC use took off worldwide with peak, annual sales of about a billion dollars (U.S.) and more than one million metric tons of CFCs produced.

Whereas CFCs are safe to use in most applications and are inert in the lower atmosphere, they do undergo significant reaction in the upper atmosphere or stratosphere. In 1974, two University of California chemists, Professor F. Sherwood Rowland and Dr. Mario Molina, showed that the CFCs could be a major source of inorganic chlorine in the stratosphere following their photolytic decomposition by UV radiation. In addition, some of the released chlorine would become active in destroying ozone in the stratosphere[2]. Ozone is a trace gas located primarily in the stratosphere (see ozone). Ozone absorbs harmful ultraviolet radiation in the wavelengths between 280 and 320 nm of the UV-B band which can cause biological damage in plants and animals. A loss of stratospheric ozone results in more harmful UV-B radiation reaching the Earth's surface. Chlorine released from CFCs destroys ozone in catalytic reactions where 100,000 molecules of ozone can be destroyed per chlorine atom.

A large springtime depletion of stratospheric ozone was getting worse each following year. This ozone loss was described in 1985 by British researcher Joe Farman and his colleagues[3]. It was called ìthe Antarctic ozone holeî by others. The ozone hole was different than ozone loss in the midlatitudes. The loss was greater over Antarctic than the midlatitudes because of many factors: the unusually cold temperatures of the region, the dynamic isolation of this ìholeî, and the synergistic reactions of chlorine and bromine[4]. Ozone loss also is enhanced in polar regions as a result of reactions involving polar stratospheric clouds (PSCs)[5] and in midlatitudes following volcanic eruptions. The need for controlling the CFCs became urgent.

In 1987, 27 nations signed a global environmental treaty, the Montreal Protocol to Reduce Substances that Deplete the Ozone Layer[6], that had a provision to reduce 1986 production levels of these compounds by 50% before the year 2000. This international agreement included restrictions on production of CFC-11, -12, -113, -114, -115, and the Halons (chemicals used as a fire extinguishing agents). An amendment approved in London in 1990 was more forceful and called for the elimination of production by the year 2000. The chlorinated solvents, methyl chloroform (CH_3CCl_3), and carbon tetrachloride (CCl_4) were added to the London Amendment.

Large amounts of reactive stratospheric chlorine in the form of chlorine monoxide (ClO) that could only result from the destruction of ozone by the CFCs in the stratosphere were observed by instruments onboard the NASA ER-2 aircraft and UARS (Upper Atmospheric Research Satellite) over some regions in North America during the winter of 1992[7,8]. The environmental concern for CFCs follows from their long atmospheric lifetime (55 years for CFC-11 and 140 years for CFC-12, CCl_2F_2)[9] which limits our ability to reduce their abundance in the atmosphere and associated future ozone loss. This resulted in the Copenhagen

Amendment that further limited production and was approved later in 1992. The manufacture of these chemicals ended for the most part on January 1, 1996. The only exceptions approved were for production within developing countries and for some exempted applications in medicine (i.e., asthma inhalators) and research. The Montreal Protocol included enforcement provisions by applying economic and trade penalties should a signatory country trade or produce these banned chemicals. A total of 148 signatory countries have now signed the Montreal Protocol. Atmospheric measurements CFC-11 and CFC-12 reported in 1993 showed that their growth rates were decreasing as result of both voluntary and mandated reductions in emissions[9]. Many CFCs and selected chlorinated solvents have either leveled off (Figure 1) or decreased in concentration by 1994[9,10].

The demand for the CFCs was accomodated by recycling, and reuse of existing stocks of CFCs and by the use of substitutes. Some applications, for example degreasing of metals and cleaning solvents for circuit boards, that once used CFCs now use halocarbon-free fluids, water (sometimes as steam), and diluted citric acids. Industry developed two classes of halocarbon substitutes- the hydrochlorofluorocarbons (HCFCs) and the hydrofluorocarbons (HFCs). The HCFCs include hydrogen atoms in addition to chlorine, fluorine, and carbon atoms. The advantage of using HCFCs is that the hydrogen reacts with tropospheric hydroxyl (OH), resulting in a shorter atmospheric lifetime. HCFC-22 ($CHClF_2$) has an atmospheric lifetime of about 13 years[11] and has been used in low-demand home air-conditioning and some refrigeration applications since 1975. However, HCFCs still contain chlorine which makes it possible for them to destroy ozone. The Copenhagen amendment calls for their production to be eliminated by the year 2030. The HFCs are considered one of the best substitutes for reducing stratospheric ozone loss because of their short lifetime and lack of chlorine. In the United States, HFC-134a is used in all new domestic automobile air conditioners. For example, HFC-134a is growing rapidly in 1995 at a growth rate of about 100% per year with an atmospheric lifetime of about 12 years[12]. (The "rule of 90" also applies for the chemical formula of HCFCs and HFCs.)

Use of the CFCs, some chlorinated solvents, and Halons should become obsolete in the next decade if the Montreal Protocol is observed by all parties and substitutes are used. The science that became the basis for the Montreal Protocol resulted in the 1995 Nobel Prize for Chemistry. The prize was awarded jointly to Professors F. S. Rowland at University of California at Irvine, M. Molina at the Massachusetts Institute of Technology, Cambridge, and Paul Crutzen at the Max-Planck-Institute for Chemistry in Mainz, Germany, for their work in atmospheric chemistry, particularly concerning the formation and decomposition of ozone (in particular, by the CFCs and oxides of nitrogen).

[1]Midgley, T., and Henne, A., Organic fluorides as refrigerants, *Industrial and Engineering Chemistry*, 22, 542-547, 1930.

[2]Molina, M.J., and F.S. Rowland, Stratospheric sink for chlorofluoromethanes: Chlorine atom catalyzed destruction of ozone, *Nature*, 249, 810-814, 1974.

[3]Farman, J.C., B.G. Gardiner, and J.D. Shanklin, Large losses of total ozone in Antarctica reveal seasonal ClOx/NOx interaction, *Nature*, 315, 207-210, 1985.

[4]McElroy, M.B., R.J. Salawitch, S.C. Wofsy, and J.A. Logan, Reductions of Antarctic ozone due to synergistic

interactions of chlorine and bromine, *Nature*, *321*, 759-762, 1986.

[5]Solomon, S., R.R. Garcia, F.S. Rowland, and D.J. Wuebbles, On the depletion of Antarctic ozone, *Nature*, *321*, 755-758, 1986.

[6]*Montreal Protocol on Substances that Deplete the Ozone Layer*, 15 pp, United Nations Environmental Programme (UNEP), New York, 1987.

[7]Toohey, D.W., L.M. Avallone, L.R. Lait, P.A. Newman, M.R. Schoeberl, D.W. Fahey, E.L. Woodbridge, and J.G. Anderson, The seasonal evolution of reactive chlorine in the northern hemisphere stratosphere, *Science*, *261*, 1134-1136, 1993.

[8]Waters, J., L. Froidevaux, W. Read, G. Manney, L. .Elson, D. Flower, R. Jarnot, and R. Harwood, Stratospheric ClO andozone from the Microwave Limb Sounder on the Upper Atmosphere Research Satellite, *Nature*, *362*, 597-602, 1993.

[9]Elkins, J.W., T.M. Thompson, T.H. Swanson, J.H. Butler, B.D. Hall, S.O. Cummings, D.A. Fisher, and A.G. Raffo, Decrease in the growth rates of atmospheric chlorofluorocarbons 11 and 12, *Nature*, *364*, 780-783, 1993.

[10]Prinn, R.G., R.F. Weiss, B.R. Miller, J. Huang, F.N. Alyea, D.M. Cunnold, P.J. Fraser, D.E. Hartley, and P.G. Simmonds, Atmospheric trends and lifetimes of CH_3CCl_3 and global OH concentrations, *Science*, *269*, 187-192, 1995.

[11]Montzka, S.A., R.C. Myers, J.H. Butler, S.C. Cummings, and J.W. Elkins, Global tropospheric distribution and calibration scale of HCFC-22, *Geophysical Research Letters*, *20* (8), 703-706, 1993.

[12]Montzka, S.A., R.C. Myers, J.H. Butler, J.W. Elkins, L.T. Lock, A.D. Clarke, and A.H. Goldstein, Observations of HFC-134a in the remote troposphere, *Geophysical Research Letters*, *23*, 169-172, 1996.

Suggested Additional Reading:

Cagin, S., and P. Dray, *Between Earth and Sky: How CFCs changed our world and threatened the ozone layer*, 512 pp., Pantheon Press, New York, 1993.

Scientific Assessment of Ozone Depletion: 1994, edited by D. L. Albritton, R. T. Watson, and R. J. Aucamp, *37*, 451 pp., World Meteorological Organization (WMO), Geneva, 1995.

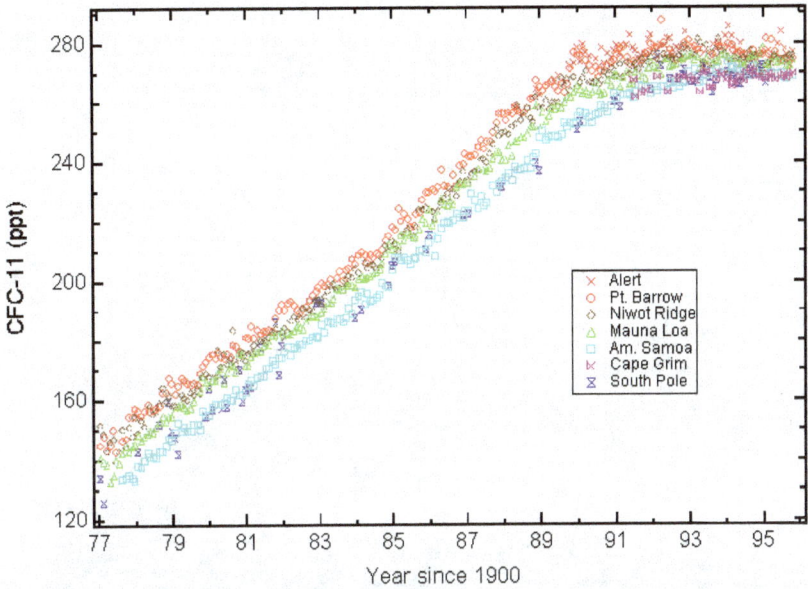

Figure 1: The accumulation of chlorofluorocarbon-11 (CFC-11) in the atmosphere levels off as a result of voluntary and mandated emission reductions. Monthly means reported as dry mixing ratios in parts per trillion (ppt) for CFC-11 at ground level for four NOAA/CMDL stations (Pt. Barrow, Alaska; Mauna Loa, Hawaii; Cape Matatula, American Samoa; and South Pole) and three cooperative stations (Alert, Northwest Territories, Canada (Atmospheric Environment Service); Niwot Ridge, Colorado (University of Colorado); Cape Grim Baseline Air Pollution Station, Tasmania, Australia, (Commonwealth Scientific and Industrial Research Organization)[9]. (Courtesy of NOAA/CMDL)

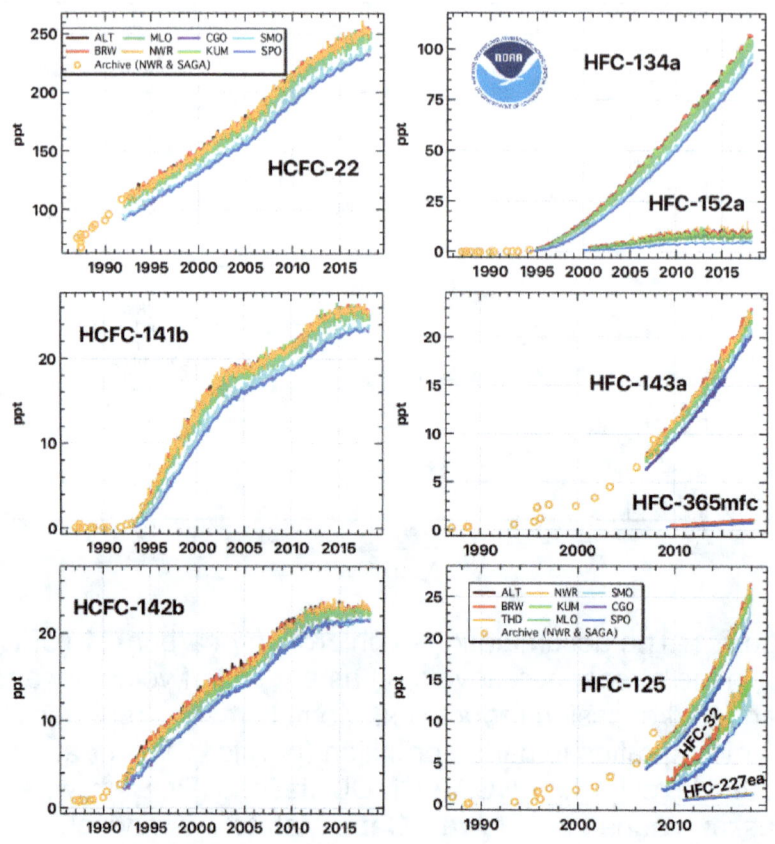

What are hydrochlorofluorocarbons (HCFCs)?

HCFCs are compounds containing carbon, hydrogen, chlorine and fluorine. Industry and the scientific community view certain chemicals within this class of compounds as acceptable temporary alternatives to chlorofluorocarbons. The HCFCs have shorter atmospheric lifetimes than CFCs and deliver less reactive chlorine to the stratosphere where the "ozone layer" is found. Consequently, it is expected that these chemicals will contribute much less to stratospheric ozone depletion than CFCs. Because they still contain chlorine and have the potential to destroy stratospheric ozone, they are viewed only as temporary replacements for the CFCs. Current international legislation has mandated production caps for HCFCs; production is prohibited after 2020 in developed countries and 2030 in developing countries.

HCFCs are less stable than CFCs because HCFC molecules contain carbon-hydrogen bonds. Hydrogen, when attached to carbon in organic compounds such as these, is attacked by the hydroxyl radical in the lower part of the atmosphere known as the troposphere. (CFCs, because they contain no hydrogen, and, therefore, no carbon-hydrogen bonds, are not destroyed by the hydroxyl radical.) When HCFCs are oxidized in the troposphere, the chlorine released typically combines with other chemicals to form compounds that dissolve in water and ice and are removed from the atmosphere by precipitation. When HCFCs become destroyed in this way their chlorine does not reach the stratosphere and contribute to ozone destruction.

A certain portion of HCFC molecules released to the atmosphere will reach the stratosphere and be destroyed there by photolysis (light-initiated decomposition). The chlorine released in the stratosphere can then participate in ozone depleting reactions as does chlorine liberated from the photolysis of CFCs. Because HCFCs are degraded significantly by two mechanisms in the atmosphere (as opposed to the CFCs which are destroyed almost exclusively by photolysis in the stratosphere), and because photolysis rates of HCFCs are generally slower than those for CFCs, proportionately less chlorine is released from HCFCs in the lower stratosphere when compared to CFCs. These properties explain why HCFCs are expected to deplete much less stratospheric ozone than equivalent amounts of CFCs.

How have atmospheric concentrations of HCFCs changed over time?

Regular, careful measurements of air from remote locations show that global concentrations of HCFCs have increased rapidly over time. This increase can be attributed to enhanced use of HCFCs as substitutes for CFCs and other chemicals as solvent/cleaning agents, refrigerants, foam-blowing agents, air conditioning fluids, etc., beginning in the late 1980s and early 1990s. Measurements of air stored in containers that were originally filled as early as 1977 and measurements of even older air trapped in snow above Antarctica or Greenland have allowed scientists at NOAA, CSIRO (Australia), and the University of East Anglia (UK) to reconstruct how concentrations of these gases have changed in the atmosphere over the past 100 years. The picture that emerges is one in which HCFCs were not present in the atmosphere in the early part of the 20th Century. Once their use was encouraged to aid in accelerating the phase out of CFCs and related ozone-depleting gases, HCFC concentrations increased rapidly from zero to the amounts observed today.

With regard to stratospheric ozone depletion, have the increases in HCFC concentrations offset the decreases observed in atmospheric concentrations of the CFCs and other ozone-depleting chemicals?

The balance of ozone in the stratosphere is determined by a number of important factors including the concentration of reactive chlorine and bromine chemicals. Dramatic increases in concentration of chlorine and bromine in the stratosphere since the 1950s have brought about enhanced ozone destruction. The destruction is most striking during the spring (Sep-Nov) over Antarctica, but is also observed to lesser extents over the Arctic during March-May and over mid-latitudes throughout the year. However, NOAA's measurements of CFCs, HCFCs and other ozone-depleting substances show that the total amount of chlorine and bromine in the atmosphere began decreasing in the 1990s! This decrease is a direct result of many nations adhering to regulations outlined in the Montreal Protocol to limit production of ozone-depleting substances. So while atmospheric concentrations of HCFCs continued to increase in the atmosphere, the declines observed to date for the more potent ozone-depleting substances (such as CFCs and methyl chloroform) have more than offset the enhanced influence of the HCFCs. Montreal Protocol regulations that restrict and eventually eliminate production of HCFCs are designed with the intent that HCFCs do not become more of a problem than the CFCs they replaced.

What do NOAA/GML measurements provide to public policy makers, the general public, and the scientific community?

NOAA/GML measurements allow the most comprehensive and consistent picture of global trace gas concentrations and changes in those concentrations over time. The trace gases we measure play an important role in determining amounts of ozone in the stratosphere. They also strongly influence the balance of heat in the atmosphere because they absorb light and because they affect other gases that absorb light (ozone in the stratosphere). Only with continued monitoring of Cl, Br, aerosols, and temperatures will we be able to answer the questions:

- Is the Montreal Protocol working?
- Is the recovery in the ozone layer proceeding as expected?
- Do any major gaps exist in our understanding of the ozone-depletion phenomena?

- Are increases in greenhouse gases delaying recovery of the ozone layer?

for more information, please contact:
Dr. Stephen A. Montzka; phone: (303)-497-6657

HCFCs are less stable than CFCs because HCFC molecules contain carbon-hydrogen bonds. Hydrogen, when attached to carbon in organic compounds such as these, is attacked by the hydroxyl radical in the lower part of the atmosphere known as the troposphere. (CFCs, because they contain no hydrogen, and, therefore, no carbon-hydrogen bonds, are not destroyed by the hydroxyl radical.) When HCFCs are oxidized in the troposphere, the chlorine released typically combines with other chemicals to form compounds that dissolve in water and ice and are removed from the atmosphere by precipitation. When HCFCs become destroyed in this way their chlorine does not reach the stratosphere and contribute to ozone destruction.

A certain portion of HCFC molecules released to the atmosphere will reach the stratosphere and be destroyed there by photolysis (light-initiated decomposition). The chlorine released in the stratosphere can then participate in ozone depleting reactions as does chlorine liberated from the photolysis of CFCs. Because HCFCs are degraded significantly by two mechanisms in the atmosphere (as opposed to the CFCs which are destroyed almost exclusively by photolysis in the stratosphere), and because photolysis rates of HCFCs are generally slower than those for CFCs, proportionately less chlorine is released from HCFCs in the lower stratosphere when compared to CFCs. These properties explain why HCFCs are expected to deplete much less stratospheric ozone than equivalent amounts of CFCs.

How have atmospheric concentrations of HCFCs changed over time?

Regular, careful measurements of air from remote locations show that global concentrations of HCFCs have increased rapidly over time. This increase can be attributed to enhanced use of HCFCs as substitutes for CFCs and other chemicals as solvent/cleaning agents, refrigerants, foam-blowing agents, air conditioning fluids, etc., beginning in the late 1980s and early 1990s. Measurements of air stored in containers that were originally filled as early as 1977 and measurements of even older air trapped in snow above Antarctica or Greenland have allowed scientists at NOAA, CSIRO (Australia), and the University of East Anglia (UK) to reconstruct how concentrations of these gases have changed in the atmosphere over the past 100 years. The picture that emerges is one in which HCFCs were not present in the atmosphere in the early part of the 20th Century. Once their use was encouraged to aid in accelerating the phase out of CFCs and related ozone-depleting gases, HCFC concentrations increased rapidly from zero to the amounts observed today.

With regard to stratospheric ozone depletion, have the increases in HCFC concentrations offset the decreases observed in atmospheric concentrations of the CFCs and other ozone-depleting chemicals?

The balance of ozone in the stratosphere is determined by a number of important factors including the concentration of reactive chlorine and bromine chemicals. Dramatic increases in concentration of chlorine and bromine in the stratosphere since the 1950s have brought about enhanced ozone destruction. The destruction is most striking during the spring (Sep-Nov) over Antarctica, but is also observed to lesser extents over the Arctic during March-May and over mid-latitudes throughout the year. However, NOAA's measurements of CFCs, HCFCs and other ozone-depleting substances show that the total amount of chlorine and bromine in the atmosphere began decreasing in the 1990s! This decrease is a direct result of many nations adhering to regulations outlined in the Montreal Protocol to limit production of ozone-depleting substances. So while atmospheric concentrations of HCFCs continued to increase in the atmosphere, the declines observed to date for the more potent ozone-depleting substances (such as CFCs and methyl chloroform) have more than offset the enhanced influence of the HCFCs. Montreal Protocol regulations that restrict and eventually eliminate production of HCFCs are designed with the intent that HCFCs do not become more of a problem than the CFCs they replaced.

What do NOAA/GML measurements provide to public policy makers, the general public, and the scientific community?

NOAA/GML measurements allow the most comprehensive and consistent picture of global trace gas concentrations and changes in those concentrations over time. The trace gases we measure play an important role in determining amounts of ozone in the stratosphere. They also strongly influence the balance of heat in the atmosphere because they absorb light and because they affect other gases that absorb light (ozone in the stratosphere). Only with continued monitoring of Cl, Br, aerosols, and temperatures will we be able to answer the questions:

- Is the Montreal Protocol working?
- Is the recovery in the ozone layer proceeding as expected?
- Do any major gaps exist in our understanding of the ozone-depletion phenomena?
- Are increases in greenhouse gases delaying recovery of the ozone layer?

for more information, please contact:
Dr. Stephen A. Montzka; phone: (303)-497-6657

Best Practices to Reduce SF6 Emissions

Reducing SF6 Emissions in Electric Power Systems: Best Industry Practices (pdf)

EPA has partnered with the electric power industry to identify and highlight cost-effective methods of reducing SF_6 emissions to the atmosphere. Utility experience shows that implementing and following best practices leads to emission reductions.

The Partnership has identified the following as best practices for reducing emissions of SF_6:

- [Establish a lifecycle approach for SF_6 management](#)
- [Establish procedures for gas inventory, accounting, and tracking](#)
- [Ensure good management of SF_6 acquisitions and gas inventory](#)
- [Train employees annually](#)
- [Recycle SF_6 gas at equipment servicing or disposal](#)
- [Implement leak detection and repair strategies](#)
- [Upgrade equipment](#)
- [Decommission equipment properly](#)

Establish a lifecycle approach for SF_6 management through company policies, protocols, and standard operating procedures. This approach helps track SF_6 inventory and costs, detect and repair leaks, and properly handle, recover, and recycle SF_6. Established policies can be continually improved upon and expanded to incorporate other options for reducing SF_6 emissions. Successful company policies and programs:

- Cover all practices
- Allow for innovation
- Designate responsible parties
- Train and empower employees

Establish procedures for gas inventory, accounting, and tracking. Tracking procedures include labeling and inventory or gas cylinders, using log sheets for warehouse cylinders, and inventorying all SF_6 equipment. Tracking procedures include labeling and inventory of gas cylinders, using log sheets for warehouse cylinders, and inventory of all SF_6 equipment. Tracking leak history of equipment identifies priorities for repairs and replacements. Tracking procedures include:

- Labeling and inventory of gas cylinders
- Using log sheets for warehouse cylinders
- Inventory of all SF_6 equipment

Ensure good management of SF_6 acquisitions and gas inventory. Utilities are consolidating storage inventory and selecting a single vendor. Vendors can support best practices by:

- Optimizing cylinder size
- Customizing the cylinder delivery system
- Minimizing cylinder handling
- Limiting inventory
- Maximizing gas utilization from every cylinder

Train employees annually in SF_6 handling and in using the necessary equipment. Training enables employees to follow procedures properly, understand the environmental and health impacts of SF_6, and learn about emission reduction options. Partners can:

- Require on-the-job training for field employees who handle SF_6
- Maintain in-house certification requirements for gas handling

Recycle SF_6 gas at equipment servicing or disposal. Using gas cart recovery equipment to off-load and transfer SF_6 for maintenance and recycling reduces emissions. It is critical to follow correct procedures when using service carts and to ensure that gas carts are properly maintained. Operators can verify that residual SF_6 is removed from equipment by:

- Using mass flow scales or weight scales
- Referring to temperature/pressure curves
- Using properly functioning recovery equipment, gauges, and scales

Implement leak detection and repair strategies. Leak detection with soap and water solutions, bagging, or thermal imaging to detect minor, chronic leaks without taking equipment out of service. Leak detection teams regularly inspect equipment to identify SF_6 leaks and prioritize repair or replacement. Technologies are available to provide real-time monitoring of SF_6 leaks and to identify components that require the most immediate repair. Leak repair is most efficient when the equipment is tested before and after repairs, using proper SF_6 recovery procedures. Effective leak repair requires advanced planning, prioritization to target worst performers, and evaluation of whether the best approach is to replace GIE.

Upgrade equipment to reduce SF_6 use and leaks. New equipment designs use less SF_6 and tighter seals to reduce leaks. Other designs use alternatives to SF_6, like vacuum-based technology with CO_2, or "Clean Air" as a base gas. While new equipment requires new maintenance procedures, training, and management adjustments, a systematic approach to anticipating equipment replacement can significantly reduce emissions. Medium-voltage alternatives have existed for the past decade; high-voltage alternatives are increasingly available.

Decommission equipment properly. Proper decommissioning using SF_6 recovery systems is important to prevent emissions. For closed-pressure systems, utilities can purify used SF_6 onsite or off-site or send non-reusable gas for destruction. Evacuate SF_6 from all equipment including hermetically sealed pressure equipment.

For more information on SF_6 emissions, please see Overview of SF_6 Emissions Sources.

Initial Response to Initial Response

Dear Mr. Diamond;

Thank you so very much for your interesting response.

Unfortunately, I continue to feel that we, (we includes the UN, EU, governments of the world and the public at large) are taking the threat presented by the super-greenhouse gas such as NF3, and a comprehensive strategy for dealing with evolving climate change much too lightly.

Thank you for including the various papers on NF3 and the graph demonstrating the annual greenhouse gas index from 1980 through 2023. However, when comparing it to the EPA's chart of U.S. Carbon Dioxide Emissions for 1990 through 2022, while the NOAA chart shows CO2 emissions increasing robustly, the EPA chart shows a decline in both gross and net CO2 emissions. (https://epa/gov/ghgemissions/inventory-us-greenhouse-gas-emissions-and-sinks

Yet, regardless of whether CO2 emissions have increased, decreased or remained steady, NF3 (as well as the other known "super greenhouse gases" remains an urgent looming environmental threat that it would be wise for mankind to address sooner rather than later.

There is little margin of error possible if we continue to put our heads in the sand when dealing with a greenhouse gas 17,200 times greater a greenhouse gas than carbon dioxide.

For example, NF3, is now widely used in the production of flat screen televisions and solar panels, both of which are experiencing substantial growth. Estimates of the annual growth in NF3 use range between 10% and 11%. Studies have also shown that more NF3 (eight times more) gets in the atmosphere than was initially thought to be the case.

I believe a much more comprehensive strategy is necessary if we are to address climate change. In your response you write: "There is certainly a concern that those gases could worsen the problem but right now they are not in the top 5 gases in our AGGI index – that does not mean that they are not important, but their contribution is relatively small at this point in comparison to big 5."

I fear, however, if we continue to ignore the problem presented by NF3 (and the other "super-greenhouse gases) and NF3 gets into the top 5, the environmental consequences will cause our current and substantial issues to pale in comparison.

There are alternatives to NF3 as well as some of the other super greenhouse gases. The international treaties and actions taken that addressed the ozone hole proved that humans can successfully address looming environmental threats.

Oft-times, however, this takes public pressure and oft-times that requires that the public be educated in the issue.

Again, thanks very much for your instructive response.

Sincerely,

R. Roy Blake

Response #2 from Howard Diamond at NOAA

Hello Richard,

Okay, thanks for that context; and I am not disagreeing with your concerns about NF3 or SF6, but all I can do is to report to you on what is going on with our monitoring. However, and I will note this, I answer an awful lot of inquiries from the public (some not very pretty) that harangue us on the issue of CO2 and convincing them that it is causing the problem that it is as the primary greenhouse gas that is leading to the radiative forcing and resultant warming that we are seeing. While I talk about greenhouse gases in general, just getting folks to focus on the chief problem is hard enough, let alone throwing things like what you are noting here; the threat from CO2 is not to minimized here, but I do not believe that we minimize the gases you note; but that said, it is a little hard to focus people's attention on gases in the parts per million range, let alone the parts per trillion range. So, I think that the book you are writing is a great thing and will hopefully elucidate the super greenhouse gases, but as I said, we are having enough trouble convincing a pretty vocal group of folks to the problem with the top 5.

So, we in NOAA are not ignoring the problem, but in fact are monitoring these gases; and if you are able to raise the attention of people to this, as well as all greenhouse gases, then that would be great. So, you are preaching to the choir here - I am all for educating the problem on all greenhouse gases and how that affects our climate.

Good luck with your efforts here.

Regards.

Howard

Initial Response #2:

From Theo Stein at NOAA

Hello Mr. Blake,

My name is Theo Stein and I'm a NOAA Communications public affairs officer who supports the Global Monitoring Laboratory in Boulder.

Your query was forwarded to me by a colleague.

The Global Monitoring Lab (GML) conducts research that addresses three major challenges, including the changing concentrations of greenhouse gases and carbon cycle feedbacks.

Specific to your question, they make highly accurate measurements of trace gases including the three primary greenhouse gases, and ozone-depleting substances, (many of which are potent greenhouse gases) from a global sampling network.

The lab and its key scientists are presently in final preparation for a five-year lab review in two weeks. This is a required **independent evaluation by external scientists of the lab's processes, scientific findings and accomplishments**. It is a key tool that NOAA uses to establish goals for its laboratory operations and programs for the next five years. A great deal of internal analysis and reporting goes into these lab reviews.

As a result, they asked me to respond to your recent query.

GML studies all influences on climate, including the carbon cycle. GML has a global sampling network and highly sophisticated labs where they analyze air samples for the most common greenhouse gases, as well as less-abundant but more potent gases.

Their Trends in CO2, CH4, N2O, SF6 provides up to date information on the four most significant GHGs.

GML's Potent GHG page describes their research monitoring the less abundant, more potent GHGs, like the ones you're interested in.

One of GML's annual reports is the Annual Greenhouse Gas Index, or the AGGI, which tracks the increasing amount of heat being added to the atmosphere by human-related greenhouse gas (GHG) emissions.

This chart from the AGGI shows the overwhelming contribution of CO_2 and methane to the greenhouse gas burden in the atmosphere. It also shows the smaller contributions of the main ozone-depleting gases, (CFCs, HCFCs, etc) which trap much more heat per molecule than CO_2, but overall trap less excess heat because CO_2 is so much more abundant. (The Montreal Protocol, designed to guide protection and recovery of the ozone layer, has had the side benefit of reducing concentrations of these potent GHGs that cause ozone depletion)

As for your question: "What steps would you recommend be taken to address the issue?"

NOAA strives to deliver accurate, authoritative, policy-relevant but policy-neutral information to decision-makers whose job it is to identify strategies for addressing greenhouse gas pollution.

While NOAA scientists do agree that to avoid the worst impacts of climate change, greenhouse gas pollution must be dramatically reduced as quickly as possible, the agency is agnostic about which policies or strategies should be pursued.

Thanks for reaching out to NOAA. I hope this information helps.

Theo Stein

Public Affairs Officer

NOAA Communications

David Skaggs Research Center

325 Broadway

Boulder, CO 80305

theo.stein@noaa.gov

303-819-7409 (cell) primary

(970) 236-8598 (Google)

Appendix I

Fighting global warming by GHG removal: Destroying CFCs and HCFCs in solar-wind power plant hybrids producing renewable energy with no-

intermittency from International Journal of Greenhouse Gas Control, Volume 49

- A large scale method of removal of GHGs other than CO_2 is proposed for CFCs & HCFs.
- Coupling two innovative and disrupting technologies allows performing GHG Removal.
- TiO_2 derivatives are possible semiconductor photocatalysts to destroy halocarbons.
- The solar-wind hybrid device + photocatalysis becomes a negative emissions technology.
- Breakthrough technologies might help to fight global warming & tackle climate change.

Abstract

In order to mitigate the effects of climate change, a lot of research has been devoted to reduce carbon dioxide (CO_2) emissions, develop capture and storage technologies (CCS), as well as direct carbon dioxide removal (CDR) from the atmosphere. The 2014 "Summary for Policymakers" of the Intergovernmental Panel on Climate Change 5th report (working group III) preponderantly mentions CCS and CDR.

Although many scientific publications cite "greenhouse gas removal", to our knowledge none of them has yet proposed innovative solutions to effectively remove greenhouse gases other than CO_2 from the atmosphere. This article proposes a combination of disrupting techniques to transform or destroy the halogenated gases in the atmosphere, which are harmful for the ozone layer, and possess high global warming potential as well as long atmospheric lifetimes.

Introduction

The anthropogenic emissions of greenhouse gases (GHGs) are primarily responsible for current global warming. Current trajectories of emissions may lead to potentially catastrophic changes in climate, as some of the available evidence suggests that scientists have been conservative in their predictions of the impacts on climate change (Brysse et al., 2013). While reduction in emissions of GHGs, particularly CO_2, could lead to a stabilization of global temperatures, it is proving difficult to curb them worldwide, as this imperatively requires international agreements which have just been reached recently at the Paris conference in December 2015.

Carbon capture and storage technologies (CCS) (Leung et al., 2014) as well as direct carbon dioxide removal (CDR) from the atmosphere (Caldecott et al., 2015) are integral parts of the IPCC (2014a) strategies to reduce the amount of CO_2 in the atmosphere.

Mitigating anthropogenic changes due to global warming will require "abatement" technologies in order to reduce GHGs emissions. But to limit the average global surface temperature increase in between 1.5 and 2 °C, removing carbon dioxide (CDR) from the atmosphere, will probably be necessary. Enhanced weathering, biochar, direct air capture, bioenergy with carbon capture and storage are some of the methods among the portfolio of CDR techniques.

If the best way to reduce global warming is by removing large volumes of CO_2 from the atmosphere, without any doubt also removing large volumes of other non-CO_2 GHGs from the atmosphere, by greenhouse gas removal (GHGR) would be even better still. The aim of this article is to propose innovative techniques for large scale atmospheric air cleaning from the halogenated gases it contains.

The Royal society (Shepherd, 2009) has evaluated the potential of different CDR technologies comparing for their effectiveness, affordability, safety and pertinence. Among them there is: afforestation (affordable but of low effectiveness); bioenergy with carbon capture; direct chemical CO_2 capture from air (effective but with low affordability), and some others. Since then many publications have completed this report from the Royal society, but to the best of our knowledge none has yet proposed innovative solutions to effectively remove greenhouse gases (GHGs) other than CO_2. Several articles mention "greenhouse gas removal" (Lomax et al., 2015), but they mainly concern a single GHG which is CO_2, or propose reducing emissions by abatement strategies, not by technological methods for direct removal from the atmosphere. This review proposes the transformation or destruction of several types of halogenated gases in the atmosphere, which are harmful for the ozone layer, possess very high global warming potential and equally, quite long atmospheric lifetimes.

After setting the goals and giving some definitions in this introductory section, in its second section this article focuses on the goal of removal or transformation of halocarbons into less harmful products for the ozone layer, and to the reduction of their global warming potential (GWP). The second part also examines the decrease of the accumulation of these halogenated gases in the atmosphere through photocatalytic processes at ambient temperature which are able to convert them into more environmentally acceptable compounds.

The third part of this review focuses on the relevance of an unusual renewable energy device (i.e. solar updraft chimney power plants (SC)) for producing decarbonized electricity.

The fourth part describes the possibilities of catching halocarbons directly from air using SCs to destroy those harmful gases.

The multiple interests of photocatalysis and the possibility of performing halons transformation, removal or destruction in synergy with solar light under the SCs are addressed in the fifth part of this review along with the possibility of a 24 h/day continuous process.

In the conclusion, the potential use of these two combined technologies and their synergy is discussed along with the potential of large-scale photocatalytic air treatments and photocatalytic applications for cleaning the atmospheric environment.

Many anthropogenic gases related to human activities (methane, nitrous oxide etc.) and also man-made industrial gases (hydrofluorocarbons, perfluorocarbons etc.) absorb infrared and, like CO_2, contribute to climate forcing. Of all GHGs, the non-CO_2 GHGs have increased antropogenic climate forcing by 45% between 1750 and 2000.

Almost all of the non-CO_2 atmospheric gases containing halogens are stratospheric ozone-depleting gases. This includes CFCs, halons, other halocarbons and some chlorinated solvents.

In this article we use halocarbons (HC) as a joint term for the halons and the other halogenated short organic molecules: chlorofluorocarbons (CFCs), hydrofluorocarbons (HFCs), perfluorocarbons (PFCs), etc. They are all long-lived well-mixed greenhouse gases (LLGHGs).

The major non-CO_2 classes of GHGs are listed by the Kyoto Protocol of the United Nations Framework Convention on Climate Change (De Boer, 2008). The Montreal Protocol for Substances that Deplete the Ozone Layer include other classes of GHGs (Anon, 2015a) like the CFCs, the halons, and the halogenated solvents containing chlorine and bromine (methyl chloroform, carbon tetrachloride, bromochloromethane, etc.).

As seen in Fig. 1, the CFCs and HCFCs contribute approximately 12% (Anon, 2015b) of the well-mixed GHGs radiative forcing (RF). The strength of different human and natural agents causing climate change is estimated by the RF, a concept used for quantitative comparisons. The influence in the Earth-atmosphere system of a GHG in altering the balance of incoming and outgoing energy is measured by the RF. The GHGs' importance and contribution to current climate change since 1750 is measured by a comparison index (expressed in W/m²) which is the global RF.

According to the National Oceanic and Atmospheric Administration (NOAA) (Anon, 2015b), CFC-12, CFC-11 and 15 other long-lived halogenated gases (CFC-113, CH_3CCl_3, CCl_4, HFCs 134a, 152a, 23, 143a, and 125, HCFCs 22, 141b and 142b, halons 1211, 1301 and 2402 and SF_6) account for about 0.282 W/m² of global RF in 2014, which is more than the global RF of nitrous oxide (0.187 W/m²), the 3rd most important climate forcer after carbon dioxide (1.909 W/m²) and methane (0.500 W/m²).

Four gases dominate the total halocarbon RF, in this order: dichlorodifluoromethane (CFC-12), trichlorofluoromethane (CFC-11), chlorodifluoromethane (HCFC-22) and trichlorofluoroethane (CFC-113). They account for about 85% of the total halocarbon RF.

In the glossary annexes of the IPCC Fifth Assessment Report: Climate Change 2013 (IPCC, 2014b) more precise definitions of RF and of global warming potential can be found.

Although CFCs emissions have been drastically reduced, their decrease takes substantial time to positively affect atmospheric concentrations because of their long lifetime. The reduction in the concentrations of CFC-11 and CFC-12 has however limited the RF from CFCs since 2005, whereas the RF from HCFCs is still rising (mainly due to HCFC-22).

For Lu, 2010, Lu, 2015, CFCs are most likely the major drivers of climate change, and not CO_2, but this theory has not yet confirmed by other researchers.

Although the emissions of several non-CO_2 GHGs seem to be easier to limit or reduce than those of CO_2, the atmospheric concentration of some of them is still rising. The concentration of HFC-134a and HFC-152a, two of the key CFC replacements, is increasing in the atmosphere. Due to this it is anticipated (see Fig. 2) that the RF of HFC-134a will be as high as or higher than the RF of all of the CFCs in the early 1990s (Anon, 2015b). Consequently, methods to reduce their atmospheric lifetime or to accelerate their natural mineralization are welcomed.

In order to remove a wide range of global warming contributors from the atmosphere, the potential applications of catalysis have been reviewed (Centi and Perathoner, 2012). However, to help fight climate change, photocatalysis seems an even more attractive method (de_Richter and Caillol, 2011). Photocatalytic processes consume less energy than thermal catalysis or other conventional methods and works by harnessing solar energy.

Tryk and Fujishima (2001) showed that *"electrochemistry can be useful in the war on global warming"* and Rajeshwar (2001) that against *"global climate change: the solar photochemical remediation approach"* using heterogeneous photocatalysis was promising.

The IUPAC 2011 definition (Braslavsky et al., 2011) of photocatalysis is a "change in the rate of a chemical reaction or its initiation under the action of ultraviolet, visible, or infrared radiation in the presence of a substance – the photocatalyst – that absorbs light and is involved in the chemical transformation of the reaction partners."

On a laboratory scale, photocatalysis has proved (Shul et al., 2010) to be a efficient method for converting many air pollutants such as volatile organic compounds (VOCs) and nitrogen oxides, to safer products for human and animal health, such as nitrates and CO_2. This will be developed further in the second part of this article.

Photocatalysis is used in multiple indoor and industrial applications to remove local air pollutants from the atmosphere. Self-cleaning glass was one of the first large scale outdoor applications of photocatalysis, followed by self-cleaning coatings and paints for buildings. The photocatalytic reduction of nitrogen oxide levels in the urban environment has been documented among the numerous large scale outdoor experiments conducted (Mo et al., 2009).

Usually at the laboratory scale, experimental photocatalytic reactions are conducted in small reactors, but outdoor applications have been conducted by applying coatings or paints containing the photocatalyst on surfaces (up to 50,000 m^2) like building walls of sidewalks in the streets. The use of solar updraft chimney power plants as giant photocatalytic reactors has been proposed (Kiesgen de_Richter et al., 2013).

A solar updraft chimney power plant (SC) is a power plant transforming the sunlight's heat collected under a greenhouse (GH), into artificial hot wind. In turn this wind drives turbines at the base of a tall tower which increases air buoyancy by a stack effect (Schlaich, 1995). This will be developed further in the third part of this article before describing the interests of coupling SC and photocatalysis with the aims of climate remediation and atmospheric air cleaning.

https://www.sciencedirect.com/science/article/abs/pii/S1750583616300858#:~:text=According%20to%20the%20National%20Oceanic,warming%20potential%20can%20be%20found

Appendix J

Best Practices to Reduce SF6 Emissions
(courtesy of NOAA)

[Reducing SF6 Emissions in Electric Power Systems: Best Industry Practices (pdf)](#)

EPA has partnered with the electric power industry to identify and highlight cost-effective methods of reducing SF_6 emissions to the atmosphere. Utility experience shows that implementing and following best practices leads to emission reductions.

The Partnership has identified the following as best practices for reducing emissions of SF_6:

- [Establish a lifecycle approach for SF_6 management](#)
- [Establish procedures for gas inventory, accounting, and tracking](#)
- [Ensure good management of SF_6 acquisitions and gas inventory](#)
- [Train employees annually](#)
- [Recycle SF_6 gas at equipment servicing or disposal](#)

- [Implement leak detection and repair strategies](#)
- [Upgrade equipment](#)
- [Decommission equipment properly](#)

Establish a lifecycle approach for SF_6 management through company policies, protocols, and standard operating procedures. This approach helps track SF_6 inventory and costs, detect and repair leaks, and properly handle, recover, and recycle SF_6. Established policies can be continually improved upon and expanded to incorporate other options for reducing SF_6 emissions. Successful company policies and programs:

- Cover all practices
- Allow for innovation
- Designate responsible parties
- Train and empower employees

Establish procedures for gas inventory, accounting, and tracking. Tracking procedures include labeling and inventory or gas cylinders, using log sheets for warehouse cylinders, and inventorying all SF_6 equipment. Tracking procedures include labeling and inventory of gas cylinders, using log sheets for warehouse cylinders, and inventory of all SF_6 equipment. Tracking leak history of equipment identifies priorities for repairs and replacements. Tracking procedures include:

- Labeling and inventory of gas cylinders
- Using log sheets for warehouse cylinders
- Inventory of all SF_6 equipment

Ensure good management of SF_6 acquisitions and gas inventory. Utilities are consolidating storage inventory and selecting a single vendor. Vendors can support best practices by:

- Optimizing cylinder size
- Customizing the cylinder delivery system
- Minimizing cylinder handling
- Limiting inventory
- Maximizing gas utilization from every cylinder

Train employees annually in SF_6 handling and in using the necessary equipment. Training enables employees to follow procedures properly, understand the environmental and health impacts of SF_6, and learn about emission reduction options. Partners can:

- Require on-the-job training for field employees who handle SF_6
- Maintain in-house certification requirements for gas handling

Recycle SF_6 gas at equipment servicing or disposal. Using gas cart recovery equipment to off-load and transfer SF_6 for maintenance and recycling reduces emissions. It is critical to follow correct procedures when using service carts and to ensure that gas carts are properly maintained. Operators can verify that residual SF_6 is removed from equipment by:

- Using mass flow scales or weight scales
- Referring to temperature/pressure curves
- Using properly functioning recovery equipment, gauges, and scales

Implement leak detection and repair strategies. Leak detection with soap and water solutions, bagging, or thermal imaging to detect minor, chronic leaks without taking equipment out of service. Leak detection teams regularly inspect equipment to identify SF_6 leaks and prioritize repair or replacement. Technologies are available to provide real-time monitoring of SF_6 leaks and to identify components that require the most immediate repair. Leak repair is most efficient when the equipment is tested before and after repairs, using proper SF_6 recovery procedures. Effective leak repair requires advanced planning, prioritization to target worst performers, and evaluation of whether the best approach is to replace GIE.

Upgrade equipment to reduce SF_6 use and leaks. New equipment designs use less SF_6 and tighter seals to reduce leaks. Other designs use alternatives to SF_6, like vacuum-based technology with CO_2, or "Clean Air" as a base gas. While new equipment requires new maintenance procedures, training, and management adjustments, a systematic approach to anticipating equipment replacement can significantly reduce emissions. Medium-voltage alternatives have existed for the past decade; high-voltage alternatives are increasingly available.

Decommission equipment properly. Proper decommissioning using SF_6 recovery systems is important to prevent emissions. For closed-pressure systems, utilities can purify used SF_6 onsite or off-site or send non-reusable gas for destruction. Evacuate SF_6 from all equipment including hermetically sealed pressure equipment.

For more information on SF_6 emissions, please see Overview of SF_6 Emissions Sources.

Electric Power Systems Partnership

- Electric Power Systems Partnership
- **Sulfur Hexafluoride (SF6) Basics**
  - Mitigation Opportunities
  - Case Studies
  - **Best Practices**
- Resources
- Events
- Connect with the EPS Partnership
- Accomplishments
-

https://www.epa.gov/eps-partnership/best-practices-reduce-sf6-emissions

Appendix K

What is Nitrogen Trifluoride (NF3)
Professor F. J. Martin-Torres
Chaired Professor in Planetary Sciences
University of Aberdeen, Scotland UK

Nitrogen trifluoride (NF3) is a synthetic inorganic chemical. It is manufactured by the reaction of hydrochloric acid (HCl) and ammonia (NH3). It is toxic, odorless, colorless, non-flammable, and is an oxidizing gaseous substance at room temperature and atmospheric pressure

1. Once NF3 is released into the atmosphere, it circulates from the surface to the stratosphere hundreds of times before it is destroyed by solar ultraviolet radiation. It is nearly chemically inert in the atmosphere, and the average lifetime of an NF3 molecule in the atmosphere is about 550 years far beyond a human lifespan.

2. NF3 is very effective in absorbing the infrared radiation that the Earth emits. By trapping this infrared radiation, NF3 becomes a potent greenhouse gas. The Global warming potential (GWP), defined as a multiple of the heat that would be absorbed by the same mass of carbon dioxide (CO_2) tells us the heat absorbed by any greenhouse gas in the atmosphere compared with CO_2 (by definition GWP is 1 for CO_2).

3. The GWPs of NF3 are 16,800, 12,200 and 18,700 for time horizons of 20-, 100-, and 500-years, respectively[2]. These values mean that NF3 is 16,800, 12,200 and 18,700 times more powerful than CO_2 in trapping atmospheric heat over a 20-, 100-, and 500-year timespan, respectively.

The increasing GWP with time horizon reflects the longer atmospheric residence time of NF3 compared with that of CO2.

4. In fact, compared to the original 6 Kyoto-listed gases, the GWP of NF3 is second only to that of SF6 (100-yr GWP = 22,800). On a per molecule basis, the radiative efficiency of NF3 is high, 0.21 W m-2 per ppb 3 , placing near the top of the low molecular weight HFCs and between CF4 (0.10) and C2F6 (0.26).

5. According to the United Nations Framework Convention on Climate Change (UNFCCC), manufacturers use NF3 as a "chamber-cleaning gas" in production processes to clean unwanted buildups on microprocessor and circuit parts as they are being constructed. A gas called hexafluoroethane (C2F6), which the first Kyoto compliance period did regulate, used to corner this market, but NF3 became a strong competitor due to its lower costs and its initial absence from the Kyoto Protocol.

6. NF3 'has a greater impact on Earth's climate per unit mass of emissions' than the C2F6 that it replaced. Industrial Applications and Increasing

Demand NF3 offers a number of advantages of relative ease of use at ambient conditions that have made it very popular in industrial applications: it is non-corrosive, non-reactive, and easy to manage chemical and reactivity hazards during storage, transportation, and normal operations.

7. It has the ability to act as a stable fluorinating agent and has a wide application scope in high-energy laser at dry etching in semiconductor production as a filling gas in lamps to prolong their durability and increase brightness, as well as a detergent gas in CVD apparatus.

8. For all these reasons, nitrogen trifluoride is increasingly used in the electronics industry, primarily for the etching of microcircuits, and for manufacturing of liquid crystal flat panel displays and thin film PV cells. The robust development of the consumer electronics market has produced a tremendous growth in the global nitrogen trifluoride market in the last years. In addition, due to the rising demand for flat panel displays, LCD televisions and several other electronic products, the market for NF3 has exploded across the globe.

9. The rising disposable income of consumers and their improving lifestyle, especially in developing economies are further estimated to accelerate the growth of the global market in the coming years. In fact, NF3 production has consequently increased 15%–17% a year, from 1,000 tons (1 Ton = 106 grams) produced in 1992 to more than 28.5 kTons (1kTon = 109 g) in 2019 (see Table 1 for the values in 2019).

10. Figure 1 shows the NF3 market volume in North America since 2012, and the current projection values for 2024, after Expert Market Research (https://www.expertmarketresearch.com/). Region Total emissions of NF3 % of worldwide market Asia 25.20 kTons 88.0% North America 2.40 kTons 8.5% Europe 1.03 kTons 3.5% Total 28.63 kTons Table 1. NF3 total emissions and % of the worldwide market in 2019. Asia is currently the largest regional NF3 market. The growth of the semiconductor industry in South Korea, China, and Japan is expected to drive augmented NF3 market demand in the production of semiconductors.

11. Demand is expected to be further propelled owing to increasing desire for flat panel displays in emerging markets such as China and India. In Europe, the increase in the government regulations to support the semiconductor and electronics industry to compete in the Asia Pacific and North American markets and to promote economic growth in the region is projected to have a positive impact on the market growth.

12. Control of NF3 emissions and legislation
According to Robson et al.3 , most of NF3 is used in a manner such that only 2% escapes to the atmosphere. Based on that study, the major NF3 producers assess that 98% of the gas is destroyed during the manufacturing process, and thus NF3 is an environmentally safe gas. However, there should be no doubt that NF3 is a highly hazardous gas in terms of global warming.

13. We have no reliable estimates about leakage during production, shipping and decommission. This gas is extremely volatile and difficult to measure at low abundances, so we cannot be sure that industry estimates or measurements are accurate. Moreover, studies by Lee et al. 4

show a maximum destruction efficiency of less than 97% under ideal conditions; and there is an economic incentive, a tradeoff between more efficient NF3 destruction and faster throughput, which encourages greater emissions. These emission fractions for NF3 do not include fugitive release during its manufacture, transport, application, or disposal. Experience with the ozone-depleting gas CFC-126, has shown that emission inventories from the chemical industry cannot be relied upon. Once released to the atmosphere, gases like CFC-12 and NF3 will take centuries to clean out. Given this potential, the production of high-GWP, long-lived, greenhouse gases like NF3 should be included in the national greenhouse gas inventories once global usage exceeds a threshold, e.g., 5 MMTCO2, no matter what the claim for containment. Returning to the intent of UNFCCC Article 3.3 ("policies and measures should... cover all relevant sources, sinks and reservoirs of green- house gases ..."), it seems prudent to expand the list of greenhouse gases to include NF3 emissions. Total NF3 emissions are significantly larger than expected assuming global implementation of ideal industrial practices. As such, there is a need for improvements in NF3 emissions reduction strategies to keep pace with its increasing use and to slow its rising contribution to anthropogenic climate forcing. NF3 is a gas of purely anthropogenic origin that has three important properties that make it a dangerous enemy: it has a large global warming potential, due to its 500 years lifetime it is accumulating in the Earth's atmosphere, and its rapid rise in production by the chemical industry has gone almost unnoticed. NF3 has a potential greenhouse impact larger than that of the industrialized nations' emissions of PFCs or SF6, or even that of the world's largest coalfired power plants. Although the instantaneous radiative forcing due to NF3 is currently small, its atmospheric lifetime essentially makes emissions cumulative with an effect lasting beyond usual societal timescales. References 1. National Center for Biotechnology Information (2020). PubChem Compound Summary for CID 24553, Nitrogen trifluoride. Retrieved October 15, 2020 from https://pubchem.ncbi.nlm.nih.gov/compound/Nitrogen-trifluoride. 2. Prather and Hsu (2008), NF3, the greenhouse gas missing from Kyoto, Geophysical Research Letters, VOL. 35, L12810, doi:10.1029/2008GL034542. 3. Robson, J. I., et al. (2006), Revised IR spectrum, radiative efficiency and global warming potential of nitrogen trifluoride, Geophys. Res. Lett., 33, L10817, doi:10.1029/2006GL026210. 4. Lee, J.-Y., et al. (2007), Evaluation method on destruction and removal efficiency of perfluorocompounds from semiconductor and display, Bull. Korean Chem. Soc., 28, 1383 – 1388. 5. Pruette, L., et al. (1999), Evaluation of a dilute nitrogen trifluoride plasmaclean in a dielectric PECVD reactor, Electrochem. Solid State Lett., 2(11),592–594. 6. Rowland, F. S. et al. (1982), Dichlorodifluoromethane, CCl2F2, in the Earth's atmosphere, Geophys. Res. Lett., 9, 481 – 484.

Appendix L

COP29 Super Pollutants (statement of the Climate Czar via the State Department to the COP29 UN Climate Conference 2024)

COP29 Super Pollutants

Super pollutants – non-CO_2 greenhouse gases like methane, nitrous oxide, and hydrofluorocarbons (HFCs) — are responsible for approximately half of near-term global warming.

(*my note: For the year 2023 the EPA says non-CO2 greenhouse gases are only responsible for 20-25% of total global warming. Either the State Department Climate Czar or the EPA are wildly wrong or the situation has changed for the worse 25-30% in just one year. Neither of those are good news*)

Tackling these potent gases is the fastest way to slow temperature rise and avoid the most severe climate impacts. The United States has prioritized tackling super pollutants through domestic regulations and international cooperation, complementing efforts to cut carbon dioxide emissions.

https://www.state.gov/cop29-releases-super-pollutants/#:~:text=Super%2Dpollutants%20%E2%80%93%20non%2DCO,the%20most%20severe%20climate%20impacts.

Appendix M

The Sprint to Cut Climate-Super-Pollutants COP29 Summit on Methane and Non-CO2 GHGS

The United States, People's Republic of China, and Azerbaijan today convened a Summit to accelerate actions to cut emissions of methane and other non-CO_2 greenhouse gases, which account for half of today's climate change but receive far less than half of global climate attention. These super pollutant greenhouse gases—including methane, hydrofluorocarbons, nitrous oxide, and tropospheric ozone—are dozens, hundreds, or even thousands of times more potent than carbon dioxide. Reducing emissions of these super pollutants is the fastest way to tackle climate change and a critical complement to reducing carbon dioxide to limit global warming to 1.5 degrees Celsius.

This year, partners announced:

- **Finance:** Over $2 billion in total international grant funding mobilized in the last three years to tackle super pollutants, and billions in additional investment deployed.
- **Policy:** New policy and regulatory steps to reduce methane emissions in the oil and gas and landfill sectors including by five of the top 20 waste sector methane emitters.
- **Science:** The launch of the first-ever Global Nitrous Oxide Assessment, which demonstrates that global N_2O emissions can be reduced by 40 percent, and new efforts to track and reduce the climate impacts of tropospheric ozone, the third-largest contributor to climate change.
-

Tackling all greenhouse gases in 2035 NDC targets and net zero targets

At the Summit, former Executive Secretary of the United Nations Framework Convention on Climate Change **Patricia Espinosa** and **International Energy Agency** Executive Director Fatih Birol called on Parties to the Paris Agreement to heed the COP 28 Global Stocktake call to deliver 1.5C-aligned 2035

Nationally Determined Contribution (NDC) targets covering all greenhouse gases, and to set net zero targets covering all greenhouse gases.

Mobilizing finance to tackle super pollutants

Building on the over $1 billion in grant funding for methane mitigation announced at COP 28 as part of President Biden's Methane Finance Sprint, multilateral development banks, philanthropy, and governments announced new finance to tackle super pollutants:

- **In total, governments and philanthropy** announced nearly $500 million in new global grant funding for methane abatement in 2024, bringing total international grant funding mobilized for super pollutant mitigation to over $2 billion in the last three years.
-
- **The European Commission** announced an additional 105.6 million Euros ($114 million) in grant funding for methane and congratulated the European Investment Bank and European Bank for

Reconstruction and Development for deploying over 2 billion Euros ($2.1 billion) and 350 million Euros ($385 million) respectively to finance methane reduction projects in the last year.

-

- **World Bank** President Ajay Banga launched a [Blueprint](#) for methane reduction at COP 28 and committed to launching 15 national programs that aim to slash methane emissions over 18 months. The World Bank Blueprint could cut up to 10 million tons of methane overall. All 15 national programs are now online, and the Bank is supporting countries through more than 40 projects to scale up methane emission reduction, with a focus on high-emitting areas like livestock, rice, waste and sanitation, and oil and gas. These projects are already helping countries to mobilize billions of dollars in International Development Association and International Bank for Reconstruction and Development financing and are supported by the Global Methane Reduction Platform for Development (CH4D) and the Global Flaring and Methane Reduction Partnership (GFMR).

-

- **International Fund for Agricultural Development (IFAD)** announced that its flagship Reducing Agricultural Methane Program (RAMP) aims to leverage up to $900 million to scale up innovative, low-methane agriculture projects. This initiative is expected to benefit over 3 million people directly and impact an additional 10 million indirectly. Additionally, IFAD introduced the Methane Reduction Guidebook for NDC 3.0, providing countries with a step-by-step guide for incorporating methane reduction strategies into their NDCs. This is complemented by targeted technical assistance to support 17 countries in designing their updated NDCs.

Enhancing national policy and ambition to cut super pollutants

At the Summit, governments announced new policy and regulatory steps to reduce super pollutants including:

- **United States** Senior Advisor to the President for International Climate Policy John Podesta announced that the United States finalized its rule to implement

the oil and gas Waste Emissions Charge which will incentivize reduction of harmful and wasteful methane pollution in the oil and gas sector. The United States also unveiled new steps to implement the oil and gas Super Emitter Program, which requires companies to take action when notified about large methane emission events. The United States released its updated **National Methane Emissions Reduction Action Plan**, which showcases more than $18 billion in funding for methane action announced and disbursed in 2024, and reinforced its intention to propose updated emissions standards for municipal solid waste landfills before the end of 2025, as required under the Clean Air Act. This year, the United States also finalized a rule to reduce leaks and promote reuse of existing HFCs, which will help the country achieve an 85% HFC phasedown by 2036. In 2024, U.S. companies also announced new actions that will reduce overall industrial emissions of N_2O by over 50% by early 2025 from 2020 levels.
- **China** Special Envoy for Climate Change Liu Zhenmin shared the actions that China has engaged in over the past year since the release of the Methane Emissions Control Action Plan, including detailing and

implementing measures, building fundamental capacities, enhancing the policy and standards framework, and actively promoting international exchange and collaboration. In particular, China is planning to lower the emission limit from the current 30 percent to 8 percent in the latest version of the *Emissions Standards of Coalbed Methane.* In addition, the Chinese government continues to improve market mechanisms, including to address methane, and Chinese enterprises are investing significantly to carry out methane emission control actions.

- **Azerbaijan** COP 29 President Mukhtar Babayev showcased the Reducing Organic Waste Declaration to catalyze urgent action to improve waste management and reduce methane emissions in the waste sector in this decade. SOCAR President Rovshan Najaf announced that SOCAR will achieve near zero methane emissions (less than 0.2 percent intensity) in upstream operations, has developed a one-stop solution for methane monitoring, and has joined the Oil and Gas Methane Partnership 2.0 (OGMP2.0).

- **Republic of Korea** Special Presidential Envoy for COP29 Hong Sik CHO launched ROK's $20 million ASEAN-Korea Cooperation on Methane Mitigation Project to support methane reduction and monitoring with ASEAN countries. ROK also advanced efforts to implement its 2030 Methane Reduction Roadmap to cut national methane emissions by over 30 percent by 2030 from 2020 levels.
- **United Arab Emirates** Ambassador Majid Al Suwaidi announced new steps to reduce methane emissions from UAE's waste sector, including setting a landfill diversion target of 50 percent diversion by 2025 and 80 percent diversion by 2031. The target will be met by implementing comprehensive national regulations to manage recycling facilities and recyclable waste and minimum fees for landfill disposal to limit random waste disposal among other measures.
- **Kazakhstan** Special Representative for International Environmental Cooperation Zulfiya Suleimenova announced that Kazakhstan will strive to complete the development of regulatory acts on methane, aiming to reduce non-emergency methane venting and promote leak detection and repair in the oil and

gas sector, as well as to finalize the National Program for Methane Emissions Reduction by COP 30 in 2025.

- **European Commission** Commissioner Hoekstra unveiled the Methane Abatement Partnership Roadmap to enhance political support for reducing emissions from internationally traded fossil fuels. The European Commission announced strengthened waste-related legislation including new measures to further reduce emissions from landfill and wastewater treatment plants. The European Commission also began implementing the EU Methane Regulation, which officially entered into force in August 2024.
- **United Kingdom** Secretary of State Ed Miliband announced £5 million ($6.5 million) of new funding in support of the Climate and Clean Air Coalition's Fossil Fuel Regulatory Program, highlighted action in the waste sector including reaffirming commitment to deliver on its flagship policy to eliminate biodegradable waste from landfill from 2028, and showcased a Methane Action Plan recently published by the Environment Agency, the main environmental regulator for England.

- **Brazil** Deputy National Secretary Aloisio Melo announced that Brazil's National Climate Change Plan will address for the first time specific targets and actions to reduce methane and other short-lived climate pollutants. Brazil also showcased its recently approved Air Quality Policy Act to update national monitoring systems and air quality standards.
- **Canada** Minister of Environment Stephan Guilbeault showcased Canada's draft regulations expected to cut landfill methane emissions approximately in half by 2030 from 2019 levels. Canada also showcased its draft regulations to cap emissions from oil and gas in upstream and liquefied natural gas production, providing an incentive for the sector to further reduce methane emissions.
- **Nigeria** Special Presidential Envoy on Climate Change Dr. Nkiruka Chidia Maduekwe showcased Nigeria's commitment to continue to champion the reduction and abatement of the super pollutants. Nigeria is collaborating with the UNEP and EU to drive the Nigeria Methane Emissions Reduction Pilot Programme (NIMERP) which is expected to expand methane abatement opportunities in the agriculture

and waste sector in support of increased ambition in NDC3.0.

Advancing new science and adopting innovative strategies for super pollutant monitoring and mitigation

Alongside the Summit, partners announced:

- **The United Nations Environment Program (UNEP) and Food and Agriculture Organization (FAO)** launched the first ever Global N_2O Assessment, followed by a commitment by the UNEP-convened Climate and Clean Air Coalition to advance scientific understanding in 2025 of opportunities to reduce the climate impacts of tropospheric ozone—an air pollutant and greenhouse gas that is formed in the atmosphere from interactions of other gases, and which is the third-largest contributor to climate change after CO_2 and methane. UNEP's International Methane Emissions Observatory intends to expand the detection component of the Methane Alert and Response System to include major methane emission events from metallurgical coal and landfills in 2025.

- **Inter IKEA Group announced** completing an analysis of their climate transition plans for sectors such as transport and energy using innovative and first-of-its-kind reporting methods for emissions of non-methane volatile organic compounds (NMVOCs), key gases that lead to the formation of the tropospheric ozone. The analysis projects the climate transition plans will result in an over 50 percent reduction in NMVOC emissions by 2030 from 2023 levels.
- **MethaneSAT and the Carbon Mapper Coalition** launched cutting-edge methane-detecting satellites this year that are already producing data and are poised to track emissions across the globe, providing the data to identify and deploy solutions, and **GHGSat announced** plans to nearly double their fleet by 2026 with the launch of nine new satellites.

For further information, please contact ClimateComms@state.gov or secretariat@ccacoalition.org.

https://www.state.gov/the-sprint-to-cut-climate-super-pollutants-cop-29-summit-on-methane-and-non-co2-ghgs/

Appendix N

COP28: What Was Achieved and What Happens Next?

The [COP 28](#) UN Climate Change Conference in Dubai, the United Arab Emirates, was the biggest of its kind. Some 85,000 participants, including more than 150 Heads of State and Government, were among the representatives of national delegations, civil society, business, Indigenous Peoples, youth, philanthropy, and international organizations in attendance at the Conference from 30 November to 13 December 2023.

[COP 28](#) was particularly momentous as it marked the conclusion of the first ['global stocktake'](#) of the world's efforts to address climate change under the [Paris Agreement](#). Having shown that progress was too slow across all areas of climate action – from reducing greenhouse gas emissions, to strengthening resilience to a changing climate, to getting the financial and technological support to vulnerable nations – countries responded with a decision on how to accelerate action across all areas by 2030. This includes a call on governments to speed up the transition away from fossil fuels to renewables such as wind and solar power in their next round of climate commitments.

Below we unpack the significance of this crucial decision and some of the key highlights from COP 28 that marked major steps forward in the global effort to address the climate emergency. Of course, the work doesn't begin and end with COP 28, so we've also outlined some of the challenges and opportunities heading into 2024 and beyond.

https://unfccc.int/cop28/5-key-takeaways

www.ingramcontent.com/pod-product-compliance
Lightning Source LLC
Chambersburg PA
CBHW071021240526
45469CB00006BD/2020